Flow Ascendancy Lift Irrigation Marvels

Flow Ascendancy Lift Irrigation Marvels

Maria .M

Mohammed Altaf Hussain

CONTENTS

INDEX

Chapter 7: Global Perspectives

7.1 Comparative analysis of lift irrigation practices in different regions

7.2 Exploration of international collaborations and knowledge exchange

7.3 Showcase of innovative approaches to lift irrigation on a global scale

Chapter 8: Future Trends and Innovations

8.1 Anticipation of emerging technologies in lift irrigation

8.2 Discussion of research and development initiatives in the field

8.3 Exploration of potential advancements and their implications for the future

Chapter 9: Conclusion

9.1 Summarization of key findings and insights

9.2 Reflection on the role of lift irrigation in sustainable agriculture

9.3 Call to action for continued research, development, and implementation of lift irrigation systems globally.

Chapter 1

Introduction

Water, the solution of life, has been a wellspring of food and thriving for civic establishments since the beginning of time. As social orders have advanced, so too have the techniques for saddling and appropriating water to meet the developing requirements of farming, industry, and day to day existence. One amazing development that has arisen in the domain of water the board is the idea of Stream Authority Lift Water system (FALI). This creative methodology tends to the difficulties of water shortage as well as achieves an extraordinary change in perspective by they way we see and use water assets.

The verifiable setting of water the executives is profoundly interlaced with the ascent and fall of civic establishments. Antiquated societies, for example, the Mesopotamians and Egyptians, perceived the significance of bridling the force of streams for water system to guarantee rural efficiency. The coming of reservoir conduits and trenches in antiquated Rome further exemplified mankind's capacity to control water stream to support society. Quick forward to the current day, and the worldwide water scene is confronting remarkable difficulties, driven by elements, for example, populace development, urbanization, environmental change, and impractical agrarian practices.

In the midst of these difficulties, Stream Command Lift Water system arises as an encouraging sign and resourcefulness. This state of the art approach use present day innovation and designing standards to improve water access, diminish energy utilization, and streamline horticultural result. The union of pressure driven designing, environmentally friendly power sources, and accuracy farming inside the system of FALI presents a comprehensive answer for the diverse water-related issues that plague locales across the globe.

To grasp the meaning of Stream Domination Lift Water system, it is basic to dig into its essential standards and the innovative wonders that support its execution. From the mechanics of lift water system frameworks to the economical practices related with FALI, this investigation will disentangle the complexities of an extraordinary methodology that can possibly reshape the eventual fate of water the board.

As we set out on this excursion, it is fundamental to perceive the more extensive setting where Stream Domination Lift Water system works. Water shortage is definitely not a restricted problem; a worldwide test requests imaginative arrangements and cooperative endeavors. FALI, with its ability to lift water from lower to higher rises, rises above geological limits, offering a plan for feasible water use that can be adjusted to different scenes.

The story unfurls against a background of expanding water pressure, where customary water system strategies miss the mark in fulfilling the needs of a blossoming populace and evolving environment. FALI remains as a demonstration of human versatility and strength, exhibiting our capacity to bridle nature's powers to defeat the restrictions forced by geography and water shortage.

The excursion into the domain of Stream Authority Lift Water system starts with an investigation of the authentic development of lift water system frameworks. Following the underlying foundations of this innovation permits us to see the value in the ever-evolving refinement of thoughts and procedures that have made ready for the modern FALI frameworks of today. From the Archimedes screw to the Persian wheel, each authentic achievement adds to the embroidered artwork of information that illuminates contemporary ways to deal with water lifting and appropriation.

As we explore through the chronicles of history, the story changes to the rise of current lift water system frameworks and the innovative headways that have molded their turn of events. The combination of power, diesel motors, and sustainable power sources has changed the effectiveness and supportability of lift water system, making way for the complex FALI frameworks that are at the very front of contemporary water the board.

The urgent job of innovation in FALI stretches out past the mechanics of water lifting; it envelops the domains of information examination, sensor advancements, and accuracy agribusiness.

The combination of these disciplines engages ranchers and water supervisors with constant bits of knowledge, empowering them to pursue informed choices that advance water utilization and harvest yields. This crossing point of innovation and farming messengers another time of brilliant water the executives, where advancement turns into the foundation of manageability.

A basic part of Stream Authority Lift Water system is its ability to outfit environmentally friendly power hotspots for controlling water lifting frameworks. The shift towards manageability isn't only a natural thought; it is a basic driven by the need to lessen functional expenses, relieve carbon impressions, and guarantee the drawn out suitability of water the executives rehearses. Sun oriented, wind, and hydropower advancements assume a crucial part in raising FALI to the situation with a green answer for water shortage, denoting a takeoff from customary energy-escalated water system techniques.

The geological pertinence of Stream Authority Lift Water system is essentially as different as the scenes it looks to change. From parched districts wrestling with intense water deficiencies to uneven landscapes where customary water system is unfeasible, FALI arises as a flexible arrangement that adjusts to the extraordinary difficulties of every climate. The contextual analyses that follow enlighten this present reality utilizations of FALI, displaying its viability in different settings and highlighting upsetting water the board on a worldwide scale potential.

Past the specialized complexities, the cultural effect of Stream Domination Lift Water system is a basic feature of its story. As water turns into an essential asset, the fair conveyance of water assets turns into a question of financial importance. FALI can possibly overcome any barrier between water-rich and water-scant locales, cultivating monetary turn of events, and relieving clashes emerging from inconsistent admittance to water. The social components of FALI reach out to enabling minimized networks, especially in provincial regions, by giving them the resources to improve farming efficiency and secure livelihoods.

In the great embroidery of water the executives, Stream Power Lift Water system arises as a string that ties together innovation, manageability, and social value. Its domination mirrors a change in perspective from conventional ways to deal with a future where water isn't simply an asset to be extricated however a unique power to be tackled and regarded. The sections that follow will dive further into the mechanics, applications, and ramifications of FALI, disentangling the layers of development that make it an extraordinary power in the mission for water security and rural supportability.

1.1 Definition of lift irrigation and its historical significance

Lift water system, at its center, is a technique for raising water from a lower to a higher height to work with the water system of horticultural grounds. This strategy remains as a demonstration of human resourcefulness in controlling normal powers to conquer geological difficulties and guarantee the effective dispersion of water assets.

The verifiable meaning of lift water system is profoundly interwoven with the advancement of civic establishments and their journey for bridling water to support agrarian efficiency.

The starting points of lift water system can be followed back to old developments where simple gadgets were utilized to raise water for water system purposes. One of the earliest and least complex types of lift water system was the Archimedes screw, ascribed to the antiquated Greek mathematician and designer Archimedes. This helical gadget, generally utilized for undertakings, for example, eliminating water from ships, was adjusted to raise water from lower to more elevated levels for water system. The Archimedes screw represents the central standards of lift water system — using mechanical power to raise water against the power of gravity.

Notwithstanding the Archimedes screw, authentic instances of lift water system incorporate the Persian wheel, otherwise called the noria. Beginning in antiquated Persia, this upward water wheel was utilized to lift water from wells or streams to flood fields. The noria was a mechanical wonder of now is the right time, driven by creature power or water flows, and its verifiable importance stretches out to locales where it was subsequently taken on, like in middle age Spain and North Africa.

The development of lift water system went on through the ages, with different societies contributing advancements to address their particular agrarian and geological difficulties. The Chinese utilized chain siphons, utilizing human or creature power, to lift water for water system. In middle age Europe, the waterwheel turned into a typical element in lift water system frameworks, driven by both even and vertical waterwheels. These authentic advancements laid the basis for the automation of lift water system in the cutting edge period.

The progress from manual and creature driven lift water system to motorized frameworks denoted a defining moment throughout the entire existence of water the board. The coming of steam motors during the Modern Transformation gave another wellspring of force for lifting water, reforming farming and empowering the development of huge breadths of beforehand parched land. The joining of steam motors into lift water system frameworks expanded productivity as well as extended the scale and extent of flooded horticulture, adding to expanded food creation and financial turn of events.

The nineteenth and mid twentieth hundreds of years saw further headways in lift water system innovation with the far reaching reception of diesel motors and electric engines. These developments gave an additional dependable and practical method for lifting water, making ready for the strengthening of horticulture in districts where water assets were a restricting variable. The motorization of lift water system expanded the size of inundated agribusiness as well as worked with the development of horticultural outskirts into beforehand undiscovered regions.

As lift water system innovation advanced, it turned into a foundation of water the executives procedures in different regions of the planet. Districts with geographical difficulties, for example, sloping territory or lopsided water dissemination, viewed lift water system as an extraordinary arrangement. The capacity to lift water from streams, lakes, or wells to higher heights considered the development of terrains that would somehow or another be unsatisfactory for conventional gravity-based water system strategies.

The authentic meaning of lift water system reaches out past its specialized angles to incorporate social, financial, and ecological aspects. In numerous social orders, the reception of lift water system frameworks assumed an essential part in supporting developing populaces by guaranteeing a steady and solid water supply for horticulture. This, thus, added to monetary flourishing and the advancement of flourishing networks revolved around watered farming.

The appearance of lift water system additionally achieved changes in land-use designs. Regions that were once thought to be unwelcoming for development because of water shortage became fruitful reason for horticulture with the presentation of lift water system. This change of scenes had significant ramifications for settlement designs, populace appropriation, and the by and large financial texture of areas that embraced lift water system advances.

In the last 50% of the twentieth 100 years, as innovation kept on propelling, lift water system frameworks developed with the mix of robotization, remote detecting, and accuracy agribusiness. These advancements worked on the productivity of water use as well as added to maintainable agrarian practices. The capacity to unequivocally control water conveyance to fields in view of continuous information introduced another period of water the executives, decreasing wastage and enhancing crop yields.

As lift water system frameworks turned out to be more refined, they likewise started to assume a critical part in water asset the executives and preservation. By empowering exact command over water circulation, these frameworks alleviated the effect of water shortage and lessen the over-extraction of groundwater assets. The mix of lift water system with current water-saving advances epitomizes the versatile limit of this well established method to address contemporary difficulties.

All in all, lift water system, with its verifiable roots tracing all the way back to old times, plays had a vital impact in forming the direction of human progress. From basic gadgets like the Archimedes screw to the perplexing and computerized frameworks of the current day, lift water system has developed because of the changing requirements of social orders. Its authentic importance lies in its commitment to horticultural advancement as well as in its extraordinary effect on scenes, settlement designs, and the financial prosperity of networks. As we dig into the contemporary period, the tradition of lift water system continues, with current cycles like Stream Authority Lift Water system (FALI) pushing the limits of development and supportability in water the board.

1.2 Overview of the importance of water management in agriculture

Water, frequently alluded to as the soul of agribusiness, is a vital asset that shapes the reasonability and maintainability of food creation frameworks around the world. The mind boggling dance among water and horticulture has been a characterizing element of human progress, and as we stand on the cliff of the 21st 100 years, the significance of successful water the board in farming has never been more articulated. This outline investigates the multi-layered meaning of water the board in farming, enveloping environmental, financial, social, and worldwide aspects.

At its center, water the executives in agribusiness includes the prudent use, dispersion, and preservation of water assets to enhance crop creation. This training is on a very basic level attached to the innate weakness of horticulture to varieties in water accessibility. Agribusiness, as an action vigorously dependent on environment examples and occasional varieties, is intensely delicate to changes in water supply, and effective water the executives turns into the key part for guaranteeing farming efficiency, food security, and natural manageability.

The environmental significance of water the executives in farming couldn't possibly be more significant. Environments are complicatedly connected to water cycles, and disturbances in water accessibility can have flowing consequences for biodiversity, soil wellbeing, and by and large biological system strength. Maintainable water the board rehearses in agribusiness look to adjust the requirements of yields with the protection of environments, limiting the unfriendly effects of water system on normal natural surroundings, oceanic frameworks, and soil quality.

With regards to water shortage and environmental change, where outrageous climate occasions and moving precipitation designs present remarkable difficulties, versatile water the board becomes basic. Effective water system frameworks, soil dampness checking, and accuracy agribusiness innovations are essential parts of this variation procedure. By advancing water use productivity, ranchers can explore the vulnerabilities of changing environments and guarantee the congruity of rural creation.

Monetarily, the significance of water the board in agribusiness is woven into the texture of worldwide food frameworks and country occupations. Horticulture is a critical supporter of public economies, and its prosperity is intrinsically attached to the accessibility and openness of water assets. Effective water the executives rehearses improve horticultural efficiency, lessen creation expenses, and increment generally speaking monetary returns for ranchers and partners along the rural worth chain.

Water system, a foundation of water the executives in horticulture, has been an extraordinary power in upgrading rural efficiency across the globe. From the old channel frameworks of Mesopotamia to the cutting edge turn and dribble water system advances, the development of water system has been inseparable from human advancement in farming.

By enhancing regular precipitation, water system permits ranchers to apply more noteworthy command over water accessibility, empowering all year development and relieving the dangers related with reliance on occasional precipitation.

The financial advantages of water system stretch out past prompt additions in crop yield. Water system works with crop expansion, permitting ranchers to develop an assortment of high-esteem crops that may not be practical under rainfed conditions. This broadening upgrades ranchers' pay as well as adds to a stronger and versatile rural area fit for enduring business sector vacillations and changing buyer requests.

Moreover, the essential utilization of water in horticulture has suggestions for worldwide food security. With a quickly developing worldwide populace and expanding urbanization, the interest for food is raising. Water-scant locales that might battle with customary rainfed horticulture can use proficient water the executives rehearses, for example, trickle water system or water gathering, to support farming result and add to meeting the world's developing food needs.

The social components of water the board in horticulture are profoundly entwined with issues of value, access, and rustic turn of events. Water shortage frequently compounds social disparities, lopsidedly influencing underestimated networks and smallholder ranchers who miss the mark on assets to execute modern water the board advancements. In numerous districts, ladies are the essential stewards of water for farming, making orientation delicate water the board significant for guaranteeing impartial admittance to this fundamental asset.

Water the executives rehearses that focus on friendly inclusivity and local area commitment add to the strengthening of rustic networks. Local area based water the board drives, for example, watershed the executives ventures or rancher cooperatives, encourage coordinated effort and aggregate direction, guaranteeing that the advantages of water use are shared impartially among local area individuals.

Also, the significance of water the executives in agribusiness reaches out to the protection of social legacy and customary cultivating rehearses. Numerous horticultural frameworks all over the planet have developed together as one with neighborhood water accessibility, epitomizing a profound comprehension of the natural subtleties of their individual scenes. Supportable water the board attempts regard and coordinate these conventional works on, recognizing the insight implanted in native information frameworks.

On a worldwide scale, the interconnectivity of water, farming, and environmental change features the direness of taking on comprehensive and feasible water the board rehearses. Horticulture is both a supporter of and a casualty of environmental change.

The discharges related with specific agrarian practices, for example, rice development and domesticated animals cultivating, add to ozone depleting substance focuses. At the same time, changing environment designs, outrageous climate occasions, and climbing temperatures present direct dangers to edit yields and water accessibility.

Practical water the executives turns into a basic methodology in relieving the effects of horticulture on environmental change and adjusting to the changing climatic circumstances. Practices, for example, agroforestry, protection horticulture, and coordinated water asset the executives add to both environment flexibility and the decrease of agribusiness' carbon impression.

All in all, the significance of water the executives in farming rises above the limits of individual fields or networks; a worldwide basic characterizes the fate of food security, ecological manageability, and country success. From the neighborhood complexities of local area based water undertakings to the general difficulties presented by environmental change, the account of water the executives in horticulture is one of transformation, development, and aggregate liability. As we explore the intricacies of a quickly influencing world, perceiving the centrality of water to the eventual fate of farming is an essential for building strong, evenhanded, and feasible food frameworks.

1.3 Introduction to the concept of "Flow Ascendancy" as a central theme

In the consistently developing scene of water the board, an idea has arisen that catches the embodiment of advancement as well as commitments a change in perspective by they way we explore the difficulties of water shortage. This idea, known as "Stream Command," remains at the convergence of water powered designing, practical horticulture, and sustainable power. It addresses an extraordinary way to deal with water lifting and circulation that goes past the customary, offering an all encompassing answer for the major problems that social orders face in dealing with their water assets. As we set out on an investigation of Stream Command, the story unfurls to uncover the standards, applications, and ramifications of this idea that holds the possibility to rethink the fate of water the board.

At its center, Stream Domination exemplifies raising water from lower to higher heights in a controlled and effective way. This climb, worked with by imaginative lift water system frameworks, challenges the limits forced by geology and water accessibility. Dissimilar to conventional strategies that depend exclusively on gravity, Stream Domination presents a dynamic and versatile aspect to water the board, engaging social orders to bridle the powers of nature for everyone's benefit.

The central standards of Stream Domination can be followed back to the old acts of lift water system, where developments cleverly controlled water through gadgets like the Archimedes screw and the Persian wheel. In any case, what separates Stream Authority is its reconciliation with present day innovation, sustainable power sources, and accuracy agribusiness. This combination denotes a takeoff from verifiable strategies, introducing a period where water the board isn't simply a utilitarian undertaking yet a many-sided hit the dance floor with nature, directed by maintainability and proficiency.

The significance of Stream Authority turns out to be especially articulated with regards to contemporary water difficulties. As populaces expand, environment designs become progressively whimsical, and regular water sources face consumption, the requirement for imaginative arrangements becomes vital. Stream Command adapts to this situation, offering a flexible and versatile methodology that can be custom-made to different scenes and water situations.

The multifaceted mechanics of Stream Domination are grounded in state of the art lift water system frameworks that use environmentally friendly power sources to hoist water. Sunlight based controlled siphons, wind turbines, and hydropower innovations become the main thrusts behind this climb, denoting a takeoff from energy-concentrated techniques for the past. The reconciliation of environmentally friendly power not just addresses the ecological effect of water lifting yet additionally adds to the more extensive worldwide shift toward manageable practices.

One of the critical features of Stream Power lies in its ability to advance water utilization through accuracy horticulture. Utilizing sensor innovations, information examination, and ongoing observing, Stream Command frameworks give ranchers bits of knowledge into soil dampness levels, crop water prerequisites, and generally field conditions. This granular way to deal with water the board guarantees that each drop is used productively, limiting waste and expanding horticultural result.

The geological flexibility of Stream Domination is an outstanding trademark that separates it from conventional water the board techniques. Bone-dry locales wrestling with intense water deficiencies, sloping landscapes where customary water system is unrealistic, and regions with lopsided water conveyance can all profit from the adaptability of Stream Command frameworks. This flexibility positions Stream Power as a flexible and versatile arrangement that can be carried out in different environments, adding to a stronger worldwide water framework.

The use of Stream Domination reaches out past the specialized domain to address cultural and financial elements of water the board. As water turns into an inexorably essential asset, the fair conveyance of water assets turns into a question of financial importance. Stream Domination can possibly overcome any issues between water-rich and water-scant locales, cultivating monetary turn of events and relieving clashes emerging from inconsistent admittance to water.

With regards to rustic turn of events, Stream Command arises as an impetus for enabling networks, especially in agrarian social orders. By giving ranchers the resources to upgrade horticultural efficiency and secure jobs, Stream Command turns into a driver of monetary development in provincial regions. The financial effect stretches out to orientation elements, with ladies frequently assuming crucial parts in water the board for agribusiness. Stream Power drives that focus on inclusivity and local area commitment can add to orientation value and strengthening.

The groundbreaking capability of Stream Command is emphasizd when seen from the perspective of worldwide supportability objectives. As countries endeavor to meet targets illustrated in drives like the Unified Countries Supportable Improvement Objectives (SDGs), Stream Authority lines up with a few key goals. From guaranteeing admittance to clean water and sterilization (SDG 6) to advancing practical horticulture (SDG 2) and reasonable and clean energy (SDG 7), Stream Command turns into a key part in the aggregate work to fabricate an additional feasible and versatile future.

The story of Stream Domination isn't restricted to hypothetical recommendations; it finds reverberation in true applications and contextual analyses. Looking at occasions where Stream Authority has been executed gives substantial proof of its viability and effect. Whether in the dry scenes of North Africa, the uneven districts of Southeast Asia, or the moving slopes of Latin America, Stream Command arises as a functional and extraordinary answer for different water the board difficulties.

As we dig into these contextual investigations, it becomes clear that the progress of Stream Command is dependent upon a conversion of elements. Mechanical development, local area commitment, strategy support, and a comprehensive comprehension of nearby environments all assume pivotal parts in deciding the viability of Stream Domination drives. These contextual investigations act as both motivation and a plan for districts looking for economical water the executives arrangements customized to their exceptional settings.

The ramifications of Stream Domination stretch out past the limits of individual tasks or districts. They echo through the worldwide talk on water the board, manageability, and environment versatility. As the worldwide local area wrestles with the interlinked difficulties of water shortage, food security, and environmental change, Stream Power offers an encouraging sign — an unmistakable appearance of humankind's capacity to enhance and adjust despite mind boggling and interconnected difficulties.

All in all, the idea of Stream Power rises above the bounds of a specialized development; it epitomizes a way of thinking that looks for concordance between human undertakings and the regular powers that oversee our planet.

As we explore the 21st hundred years, where the crossing points of populace development, environmental change, and asset exhaustion combine, the significance of creative water the executives rehearses couldn't possibly be more significant. Stream Power remains as a demonstration of our ability to rethink the manner in which we connect with water, transforming the difficulties of shortage into valuable open doors for reasonable development and aggregate prosperity. The parts that follow will dive further into the mechanics, applications, and cultural ramifications of Stream Domination, disentangling the layers of development that make it a groundbreaking power in the mission for water security and farming maintainability.

In the unpredictable embroidery of water the board, the idea of "Stream Power" arises as a focal topic, offering a progressive point of view on how we tackle and disseminate water assets. This imaginative idea goes past traditional strategies, presenting a dynamic and reasonable way to deal with raising water from lower to higher rises. Grounded in pressure driven designing, environmentally friendly power, and accuracy farming, Stream Domination addresses a change in outlook in the domain of water lifting, holding the possibility to reshape the eventual fate of water the board. As we explore through the standards, applications, and ramifications of Stream Power, a thorough comprehension of this extraordinary idea unfurls.

At its substance, Stream Power typifies the most common way of lifting water in a controlled and productive way, testing the limitations forced by geography and customary water system strategies. The verifiable precursors of this idea can be followed back to old human advancements where simple gadgets like the Archimedes screw and the Persian wheel were utilized to hoist water for agrarian purposes. Notwithstanding, what recognizes Stream Command is its contemporary mix with trend setting innovation, sustainable power sources, and accuracy horticulture rehearses.

The center standards of Stream Domination lie in its takeoff from the dependence on gravity alone, presenting a powerful exchange between creative lift water system frameworks and sustainable power. This reconciliation denotes a takeoff from verifiable practices and lines up with the worldwide shift towards feasible and eco-accommodating innovations. The accentuation on sustainable power sources, for example, sunlight based, wind, and hydropower not just addresses the ecological effect of water lifting yet in addition mirrors a pledge to lessening carbon impressions and cultivating long haul maintainability.

The mechanics of Stream Authority are exemplified in current lift water system frameworks that influence environmentally friendly power for water height. Sunlight based fueled siphons, wind turbines, and limited scope hydropower units assume a urgent part in this cycle, changing the demonstration of lifting water into a more maintainable and energy-productive undertaking. This takeoff from ordinary, energy-serious techniques implies a cognizant work to adjust water the executives practices to more extensive worldwide objectives of relieving environmental change and progressing towards cleaner, greener innovations.

The flexibility of Stream Power is a central trait that separates it from conventional water the executives strategies. This idea isn't restricted to explicit geological or geographical requirements; rather, it flourishes in assorted scenes, taking special care of the remarkable difficulties presented by dry locales, bumpy landscapes, or regions with lopsided water appropriation. The flexibility of Stream Power positions it as a versatile arrangement that can be modified to suit the fluctuated needs of various environments, adding to a stronger and adaptable worldwide water foundation.

Accuracy farming, a vital part of Stream Command, includes the utilization of sensor innovations, information investigation, and ongoing observing to improve water use effectiveness. By furnishing ranchers with granular experiences into soil dampness levels, crop water prerequisites, and generally speaking field conditions, accuracy agribusiness guarantees that water is used prudently, limiting wastage and boosting agrarian result. This reconciliation of innovation and farming mirrors a pledge to shrewd, information driven dynamic in the domain of water the executives.

As we dig further into the uses of Stream Domination, the meaning of this idea in tending to worldwide water difficulties turns out to be progressively obvious. In locales where water shortage is a major problem, Stream Power offers a life saver by empowering the development of parched lands and improving horticultural efficiency. The capacity to lift water to higher rises extends the potential for water system, changing scenes that were once thought to be unsatisfactory for conventional gravity-based techniques.

Contextual analyses from around the world act as demonstrations of the viability of Stream Command in different settings. In bone-dry locales of North Africa, where water shortage represents a critical danger to farming, Stream Domination frameworks controlled by sun based energy have shown their capacity to reasonably hoist water for water system. Likewise, in the rugged landscapes of Southeast Asia, where traditional water system is frequently unrealistic, Stream Command arises as a feasible answer for conquer geographical difficulties and guarantee water arrives at farming fields effectively.

Latin America, with its assorted scenes, presents one more material for the use of Stream Domination. From the huge fields to the moving slopes, the versatility of this idea becomes obvious. In locales with lopsided water conveyance, Stream Command offers a way to lift water from accessible sources to raised horticultural terrains, guaranteeing a more impartial circulation of this valuable asset. These contextual analyses highlight the flexibility of Stream Authority in tending to explicit difficulties inside exceptional settings.

The cultural ramifications of Stream Authority stretch out past mechanical advancement to incorporate monetary turn of events, social value, and country strengthening. As water turns into an essential asset, the evenhanded dissemination of water assets arises as a financial goal.

Stream Command can possibly overcome any barrier between water-rich and water-scant districts, cultivating monetary turn of events and moderating contentions emerging from inconsistent admittance to water.

In the domain of country improvement, Stream Domination arises as an impetus for enabling networks, especially in agrarian social orders. By giving ranchers the necessary resources to upgrade horticultural efficiency and secure occupations, Stream Power turns into a driver of monetary development in country regions. The financial effect reaches out to orientation elements, with ladies frequently assuming crucial parts in water the board for agribusiness. Stream Authority drives that focus on inclusivity and local area commitment can add to orientation value and strengthening.

On a worldwide scale, Stream Domination lines up with the standards and targets framed in global drives like the Unified Countries Economical Improvement Objectives (SDGs). From guaranteeing admittance to clean water and sterilization (SDG 6) to advancing manageable horticulture (SDG 2) and reasonable and clean energy (SDG 7), Stream Power turns into a key part in the aggregate work to construct an additional maintainable and strong future.

The ramifications of Stream Command reach out past the prompt domains of individual activities or districts; they resonate through the worldwide talk on water the executives, manageability, and environment flexibility. In a world wrestling with interconnected difficulties of water shortage, food security, and environmental change, Stream Command offers a substantial sign of humankind's capacity to enhance and adjust.

All in all, Stream Domination arises as an extraordinary idea that rises above conventional ways to deal with water the board. It encapsulates a way of thinking that looks for concordance between human undertakings and the normal powers that oversee our planet. As we explore the intricacies of the 21st 100 years, where the crossing points of populace development, environmental change, and asset consumption meet, the significance of creative water the board rehearses couldn't possibly be more significant. Stream Command remains as a demonstration of our ability to rethink the manner in which we collaborate with water, transforming the difficulties of shortage into open doors for maintainable development and aggregate prosperity. The resulting investigation in the parts that follow will dive further into the mechanics, applications, and cultural ramifications of Stream Domination, disentangling the layers of development that make it an extraordinary power in the mission for water security and rural manageability.

Chapter 2

The Genesis of Lift Irrigation

The beginning of lift water system can be followed back to the antiquated civilizations that prospered along the banks of streams. These civic establishments, like the Indus Valley Civilization in South Asia, depended vigorously on the accessibility of water for their food and farming exercises. The brilliant utilization of straightforward water system procedures permitted these antiquated social orders to bridle the force of water to develop their harvests and backing their developing populaces.

One of the earliest types of water system included redirecting water from streams or other water bodies to fields through channels or trenches. This gravity-based water system technique was powerful partially, however it had its constraints. The accessibility of water was dependent upon the normal progression of streams, and ranchers had little command over the timing and amount of water arriving at their fields.

As social orders advanced and urbanized, the interest for water expanded, and the requirement for more refined water system frameworks became evident. The idea of lift water system started to come to fruition as an answer for the difficulties presented by conventional gravity-based frameworks. Lift water system includes raising water from a lower rise to a higher height, permitting ranchers to inundate their fields regardless of whether they are arranged at a higher height than the water source.

The coming of mechanical gadgets denoted a huge achievement in the development of lift water system. Early civilizations created basic gadgets, for example, the Archimedean screw and water wheels to lift water from streams or wells to higher heights. These gadgets, however simple by the present norms, established the groundwork for the further developed lift water system frameworks that would arise in the hundreds of years to come.

In the archaic period, different societies across the world proceeded to refine and further develop lift water system advancements. In Persia, for instance, the Shaduf — a hand-worked gadget comprising of a switch and a can — was broadly utilized for lifting water from wells or trenches. The Shaduf permitted ranchers to physically lift water to flood their fields, giving a level of command over water dissemination.

In South Asia, especially in areas where the storm downpours were not adequate for all year agribusiness, old civilizations created imaginative procedures to outfit water for water system. Stepwells, known as "baolis" in India, were developed to store water and give a solid wellspring of water during dry seasons. These stepwells filled a double need of water stockpiling and lifting water to more elevated levels for farming use.

The Renaissance time frame in Europe saw a resurgence of interest in science and designing, prompting further progressions in lift water system innovation. Designers and creators of the time started exploring different avenues regarding different mechanical gadgets to productively lift water more. The presentation of pinion wheels, wrenches, and pulleys took into consideration the advancement of more complex and strong water-lifting components.

The Modern Upheaval in the eighteenth and nineteenth hundreds of years denoted a defining moment throughout the entire existence of lift water system. The creation of steam motors and the outfitting of steam power upset the manner in which water could be lifted. Steam-driven siphons and motors were utilized to lift water from streams, lakes, or wells to higher heights, empowering huge scope water system projects that were unrealistic previously.

Provincial powers, as they continued looking for horticultural development and monetary predominance, assumed a huge part in presenting and carrying out lift water system frameworks in different regions of the planet. The English, for instance, embraced aggressive water system projects in their provinces, for example, the development of channels and lift frameworks to work with horticulture in bone-dry locales.

The twentieth century saw further mechanical developments in lift water system, with the boundless reception of electric siphons and engines. Electrically controlled lift frameworks gave an additional solid and effective method for lifting water, lessening reliance on physical work and steam power. This mechanical jump fundamentally expanded the practicality of lift water system for a more extensive scope of farming applications.

The mid-twentieth century saw the rise of huge scope lift water system projects, especially in agricultural nations confronting water shortage challenges. Legislatures and worldwide associations put resources into foundation to bridle water assets for rural turn of events. The Green Upset, which intended to increment food creation through the presentation of high-yielding harvest assortments, was firmly connected to the development of lift water system frameworks.

Lift water system innovation kept on developing with the approach of computerization and mechanization in the last 50% of the twentieth hundred years. Mechanized control frameworks empowered exact administration of water stream, improving water system effectiveness and limiting water wastage. Remote detecting advances and information examination further upgraded the checking and the board of lift water system frameworks, introducing another period of shrewd farming.

In the 21st 100 years, the difficulties of environmental change, populace development, and expanding water shortage have moved the improvement of reasonable and eco-accommodating lift water system arrangements. Developments, for example, sun oriented fueled siphons and energy-proficient lift frameworks have acquired unmistakable quality as the world looks to change to additional harmless to the ecosystem rural practices.

The beginning of lift water system, established in antiquated civilizations and molded by hundreds of years of mechanical progressions, keeps on developing because of the squeezing difficulties of the contemporary world. The capacity to lift water to higher heights has worked with horticultural improvement as well as assumed a vital part in guaranteeing food security and monetary flourishing for networks all over the planet.

In spite of its verifiable importance and the extraordinary effect it has had on farming, lift water system isn't without its contentions and difficulties. The huge scope redirection of water for water system purposes has raised worries about the consumption of regular water sources, natural uneven characters, and clashes over water privileges. Finding some kind of harmony between the requirement for water system and the feasible administration of water assets stays a mind boggling and continuous undertaking for policymakers, designers, and networks depending on lift water system.

As we plan ahead, the proceeded with refinement of lift water system innovation will probably be formed by a blend of variables, remembering progressions for sustainable power, accuracy farming, and water the executives rehearses. The coordination of present day sensor advancements, computerized reasoning, and information driven dynamic will add to the improvement of more productive and manageable lift water system frameworks.

All in all, the beginning of lift water system is a demonstration of humankind's creativity and flexibility in saddling the fundamental asset of water for farming flourishing. From antiquated civic establishments depending on straightforward gadgets to the modern, innovation driven frameworks of today, the excursion of lift water system mirrors the unique exchange between human development and the difficulties presented by the consistently changing horticultural scene. As we stand at the junction of a quickly advancing world, the historical backdrop of lift water system gives significant experiences into the complicated connection between water, horticulture, and the economical improvement of social orders.

2.1 Historical development of lift irrigation techniques

The authentic advancement of lift water system methods traverses centuries, mirroring the developing necessities of social orders and the resourcefulness of human civic establishments in saddling water for horticulture. This excursion starts in antiquated times, where early rural networks looked for ways of conquering the impediments of depending exclusively on regular water sources.

In the support of antiquated civic establishments, along the banks of streams like the Indus, Nile, and Tigris-Euphrates, individuals perceived the basic job of water in horticulture. The rise of early water system methods established the groundwork for the improvement of lift water system. Basic strategies, for example, redirecting water from streams into fields through channels or trenches, permitted ranchers to control water stream somewhat. Notwithstanding, these gravity-based frameworks had innate restrictions, as water accessibility was dependent upon the normal flow of streams.

As social orders progressed, so did the requirement for more modern water system techniques. The idea of lifting water from lower to higher rises started to come to fruition. In old China, the utilization of a chain siphon, credited to the designer Du Shi during the Han Tradition (around first century Promotion), showed an early type of lift water system. The chain siphon, worked by difficult work, been able to raise water effectively, denoting a mechanical jump in water system rehearses.

Likewise, antiquated human advancements in South Asia formulated creative strategies to address water shortage and rise difficulties. Stepwells, for example, the renowned Chand Baori in Rajasthan, India, filled in as both water stockpiling designs and gadgets to lift water to more elevated levels. These stepwells were compositional wonders, highlighting complicated flights of stairs driving down to the water level and working with the extraction of water for agrarian use.

The archaic period saw the refinement of lift water system strategies in different areas of the planet. In Persia, the Shaduf, a hand-worked gadget with a switch and a can, became broad. Ranchers utilized the Shaduf to physically lift water from wells or channels, giving a level of command over water conveyance. These manual gadgets were urgent for networks that required a dependable water supply for their harvests.

The Renaissance time frame in Europe denoted a resurgence of interest in science and designing. Designers and creators investigated ways of improving water system methods. Gadgets like the Archimedean screw and water wheels acquired notoriety, considering more productive lifting of water. The utilization of cog wheels, wrenches, and pulleys in these systems addressed a stage towards motorization in lift water system.

The Modern Upheaval in the eighteenth and nineteenth hundreds of years proclaimed a huge change in lift water system innovation. Steam motors, a sign of this period, reformed water lifting. Steam-driven siphons and motors could lift water from streams, lakes, or wells to higher heights, empowering enormous scope water system projects. Pilgrim powers, driven by rural development and financial interests, carried out aggressive lift water system projects in their settlements, leaving an enduring effect on water system framework.

As the twentieth century unfurled, lift water system kept on advancing with the reception of electric power. Electric siphons and engines supplanted steam motors, offering an additional dependable and effective method for lifting water. This change from steam to electric power fundamentally expanded the practicality of lift water system, making it open to a more extensive scope of rural applications.

The mid-twentieth century saw the rise of enormous scope lift water system projects, particularly in non-industrial nations confronting water shortage challenges. State run administrations and worldwide associations put resources into water system framework to outfit water assets for farming turn of events. The Green Unrest, a worldwide drive to increment food creation, was firmly connected to the extension of lift water system frameworks. High-yielding harvest assortments required dependable and controlled water sources, provoking the execution of refined water system methods.

Progressions in lift water system innovation went on with the approach of computerization and mechanization in the last 50% of the twentieth hundred years. Computerized control frameworks took into consideration exact administration of water stream, enhancing water system proficiency and limiting water wastage. Remote detecting innovations and information examination further improved the observing and the executives of lift water system frameworks, introducing another period of brilliant farming.

In the 21st hundred years, lift water system faces new difficulties presented by environmental change, populace development, and expanding water shortage. Imaginative and manageable arrangements are tried to address these difficulties. Sun oriented controlled siphons, tackling energy from the sun to lift water, have acquired unmistakable quality as a component of harmless to the ecosystem farming practices. Energy-effective lift frameworks and accuracy farming advances add to the mission for supportability in water use.

Regardless of the innovative headways and the positive effect of lift water system on agrarian efficiency, debates and difficulties persevere. The enormous scope redirection of water for water system raises worries about the consumption of normal water sources, biological uneven characters, and clashes over water freedoms. Finding some kind of harmony between the requirement for water system and the feasible administration of water assets stays a complicated and progressing task for policymakers, designers, and networks dependent on lift water system.

Looking forward, the direction of lift water system improvement is probably going to be impacted by a mix of variables, remembering headways for sustainable power, accuracy farming, and water the board rehearses. Incorporation of current sensor advancements, man-made reasoning, and information driven dynamic will add to the improvement of more proficient and supportable lift water system frameworks.

All in all, the authentic improvement of lift water system procedures mirrors the ceaseless journey of human social orders to conquer the difficulties presented by water shortage and rise differentials in agribusiness. From antiquated manual gadgets to complex, automated frameworks of the current day, the advancement of lift water system reflects the unique connection between human development and the basic to guarantee food security through effective water use. As we stand at the crossing point of innovative advancement and natural awareness, the historical backdrop of lift water system gives significant bits of knowledge into the mind boggling interchange between water, horticulture, and the supportable improvement of social orders.

2.2 Early innovations and their impact on agricultural practices

Early developments play had a urgent impact in forming horticultural practices all through mankind's set of experiences. As social orders changed from traveling ways of life to settled rural networks, shrewd arrangements arose to improve efficiency, guarantee food security, and adjust to the difficulties presented by different conditions. This investigation starts with the Neolithic Upheaval, a groundbreaking period that saw the coming of horticulture.

The Neolithic Transformation, happening around 10,000 BCE, denoted a significant change in mankind's set of experiences as individuals progressed from hunting and assembling to farming. This progress was driven by the development of plants and the training of creatures.

In the Prolific Sickle, a region enveloping pieces of advanced Iraq, Iran, Syria, and Turkey, early ranchers developed wheat and grain and trained animals like goats and sheep. The advancement of basic instruments, like stone sickles and scrapers, worked with the plowing of soil and planting of seeds.

The advancement of horticulture achieved tremendous changes in settlement designs. People started to lay out long-lasting towns, taking into account the improvement of additional perplexing social designs. The capacity to create an overflow of food empowered populace development and the rise of particular jobs inside networks. These early rural advancements established the groundwork for the foundation of civic establishments, molding the course of human turn of events.

In old Mesopotamia, the support of progress, complex water system frameworks were created to outfit the waters of the Tigris and Euphrates streams. The development of trenches and levees took into account controlled water dissemination, guaranteeing predictable water system for crops. The Balancing Nurseries of Babylon, one of the Seven Miracles of the Old World, is a demonstration of the dominance of water system and cultivation around here.

The antiquated Egyptians also succeeded in farming development along the Nile Stream. The yearly flooding of the Nile brought supplement rich dregs, making fruitful soil for development. Egyptians fostered an arrangement of bowl water system, redirecting water from the Nile into bowls and permitting it to flood fields. The Shaduf, a hand-worked gadget with a switch and stabilizer, was utilized for lifting water from the Nile to flood higher fields.

In China, old horticultural advancements were driven by the need to support enormous populaces. The development of the Fabulous Trench during the Sui and Tang traditions worked with the transportation of merchandise and considered broad rice development. Patio cultivating, particularly in hilly locales, limited soil disintegration and augmented arable land. The utilization of water wheels for water system displayed China's initial dominance of pressure driven designing.

The Greco-Roman world likewise seen developments in horticulture. The Roman Domain, with its tremendous regions, fostered a broad organization of reservoir conduits to move water for farming and metropolitan use. Pliny the Senior, a Roman naturalist, recorded rural practices in his work "Naturalis Historia," giving experiences into the information and strategies utilized by Roman ranchers.

Middle age Europe saw the presentation of the three-field framework, a rotational trimming strategy that expanded soil richness and harvest yields. The utilization of iron furrows supplanted wooden furrows, taking into account more productive development. Windmills and watermills arose as wellsprings of mechanical power for errands like crushing grain and emptying water out of low-lying fields.

The Islamic Brilliant Age (eighth to fourteenth hundreds of years) contributed essentially to rural information. Islamic researchers deciphered antiquated Greek and Roman texts, safeguarding and developing horticultural insight. Developments, for example, the Noria, a water wheel utilized for water system, exemplified the mechanical headways of this time.

In the Americas, native civilizations created refined horticultural practices. The Maya, Aztec, and Inca developments developed crops like maize, beans, and potatoes. The Inca, specifically, designed terraced fields on the precarious slants of the Andes, boosting arable land. The chinampas, counterfeit islands utilized for farming, displayed the resourcefulness of the Aztecs in overseeing water assets.

As social orders entered the Renaissance period in Europe, a reestablished interest in science and investigation impacted farming practices. The presentation of new yields, like potatoes and maize from the Americas, broadened European weight control plans and further developed sustenance. The agrarian upheaval in the eighteenth century saw the motorization of cultivating with developments like the seed drill and the walled in area development, which combined land for more productive development.

The nineteenth century achieved further advancements in farming with the improvement of the steel furrow by John Deere and the collector by Cyrus McCormick. These creations upset the productivity of hands on work, permitting ranchers to develop bigger regions and gather crops all the more quickly. The spread of railways and further developed transportation worked with the conveyance of horticultural items to far off business sectors.

The twentieth century denoted a time of remarkable mechanical headway in farming, known as the Green Unrest. Researchers like Norman Borlaug presented high-yielding assortments of harvests, alongside the utilization of manufactured composts and pesticides. This upset decisively expanded crop yields and turned away starvations in many regions of the planet. Apparatus, for example, work vehicles and consolidate gatherers became ordinary on ranches, changing the scale and proficiency of horticulture.

The last 50% of the twentieth century saw the rise of accuracy farming, empowered by advancements like worldwide situating frameworks (GPS) and remote detecting. Ranchers could now enhance planting, water system, and reaping in view of itemized information, prompting asset productivity and diminished natural effect. Biotechnology presented hereditarily changed crops, further upgrading protection from nuisances and infections.

Lately, the attention on supportability has driven horticultural advancement. Rehearses like natural cultivating, agroforestry, and regenerative agribusiness mean to advance biological equilibrium and limit ecological effect. The coordination of computerized advances, including the Web of Things (IoT) and man-made consciousness, considers ongoing observing and dynamic in cultivating tasks.

Vertical cultivating and aquaculture address advanced ways to deal with farming, testing customary ideas of developing yields in soil. These techniques, which include developing plants in controlled conditions with negligible water utilization, offer likely answers for issues of land shortage and water deficiencies.

The effect of early advancements on agrarian practices is significant and expansive. The progress from agrarian social orders to settled farming laid the basis for civilization. Antiquated water system frameworks empowered the development of harvests in parched locales, prompting the foundation of cutting edge social orders. Developments in apparatuses, crop turn, and water system strategies during the middle age time frame expanded horticultural efficiency and manageability.

The agrarian information on the Islamic Brilliant Age and the commitments of native civilizations in the Americas added assorted points of view to worldwide cultivating rehearses. The Renaissance and ensuing horticultural transformations introduced a time of automation and expanded effectiveness. The Green Upset in the twentieth century changed worldwide farming, tending to hunger and expanding food creation.

The combination of innovation in the 21st century keeps on rethinking agribusiness, offering answers for ecological difficulties and expanding the accuracy and manageability of cultivating rehearses. As humankind wrestles with the intricacies of taking care of a developing populace while safeguarding the planet, the examples gained from early advancements give significant experiences into the flexibility and versatility of farming practices throughout the span of history.

2.3 Evolution of lift irrigation systems across different cultures and civilizations

The development of lift water system frameworks across various societies and civilizations is a demonstration of human inventiveness and the tenacious quest for proficient water the executives for farming purposes. From old manual gadgets to current, innovatively progressed frameworks, the improvement of lift water system has been molded by the novel geological, climatic, and cultural difficulties looked by different developments over the entire course of time.

In the chronicles of old civic establishments, the usage of lift water system frameworks was critical for supporting horticulture in districts where normal water sources were not effectively available or where height differentials presented difficulties. The Indus Valley Civilization, one of the world's most seasoned metropolitan societies, gives early proof of refined water the board. The occupants of Mohenjo-daro and Harappa, the significant urban areas of the human advancement, had an intricate arrangement of underground depletes and channels for wastewater removal and possibly water system. While the particulars of lift water system in the Indus Valley stay speculative, the dominance of water designing was clear in their metropolitan preparation.

In old Egypt, the yearly flooding of the Nile Stream assumed a focal part in the farming cycle. Nonetheless, in regions past the compass of the flooding, lift water system procedures were utilized. The Shaduf, a basic hand-worked gadget with a long post and stabilizer, permitted Egyptian ranchers to lift water from the Nile to inundate fields arranged at higher heights. This manual framework displayed the genius of antiquated Egyptians in streamlining water use for farming.

The antiquated Persians, famous for their commitments to designing and water the executives, carried out different lift water system procedures. The Qanat framework, an underground passage that took advantage of springs and moved water to the surface, filled in as a creative technique for lifting water to higher rises. Furthermore, the Shaduf, like its utilization in Egypt, tracked down application in Persian farming, featuring the diverse dispersion of lift water system advancements.

In old China, where farming practices were fundamental to the human progress' turn of events, the idea of water lifting was exemplified in gadgets like the chain siphon. This early mechanical development, credited to design Du Shi during the Han Tradition (around first century Promotion), involved a consistent chain of pails that lifted water from a lower to a higher rise. While at first worked physically, the chain siphon laid the preparation for more refined lift water system frameworks in later hundreds of years.

The Greco-Roman world additionally seen headways in water-lifting innovation. Archimedes, the old Greek mathematician and creator, is credited with formulating the Archimedean screw — a helical siphon that could lift water from a lower to a more elevated level. This screw instrument turned into a central part of different lift water system frameworks across various societies and time spans. Additionally, water wheels, which tackled the force of streaming water to lift it to higher heights, were used by antiquated Romans in specific rural settings.

As the archaic period unfurled, the Islamic Brilliant Age saw huge commitments to pressure driven designing and lift water system. The Noria, an enormous water wheel frequently connected with Islamic civilizations, exemplified the innovative ability of the time. Norias were utilized to lift water from streams or wells to inundate fields, and their plan shifted across locales. These water wheels became notorious images of Islamic advancement in water the executives.

In South Asia, especially in India, the stepwells known as "baolis" displayed a remarkable way to deal with lift water system. These stepwells, like the Chand Baori in Rajasthan, were engineering wonders as well as filled the double need of water stockpiling and lifting water to more significant levels. The complicated plan of steps worked with admittance to water at various levels, taking into account proficient water extraction and appropriation.

The Renaissance time frame in Europe saw a resurgence of interest in science and designing, prompting further developments in lift water system innovation. Specialists and creators investigated different mechanical gadgets to productively lift water more. Water wheels, propelled by antiquated plans, were refined and utilized in different settings, adding to horticultural turn of events.

The Modern Upset in the eighteenth and nineteenth hundreds of years denoted an extraordinary stage in the development of lift water system. The development of steam motors and the saddling of steam power upset water-lifting components. Steam-driven siphons and motors could lift water from streams or wells to higher rises, empowering huge scope water system projects that were not attainable with manual or creature fueled frameworks.

Pioneer powers, driven by financial interests and the longing for rural extension, assumed a huge part in presenting and carrying out lift water system frameworks in different regions of the planet. The English, for example, embraced aggressive water system projects in their states, developing channels and lift frameworks to work with horticulture in dry districts. This provincial impact left an enduring engraving on the water system foundation of numerous locales.

The twentieth century saw a change in perspective in lift water system with the far and wide reception of electric power. Electric siphons and engines supplanted steam motors, giving an additional dependable and productive method for lifting water. This change essentially expanded the plausibility of lift water system, making it open for a more extensive scope of rural applications.

Enormous scope lift water system projects became noticeable during the twentieth hundred years, particularly in emerging nations confronting water shortage challenges. Legislatures and global associations put resources into foundation to tackle water assets for farming turn of events. The Green Transformation, a worldwide drive to increment food creation, was firmly connected to the extension of lift water system frameworks. High-yielding harvest assortments required controlled and solid water sources, driving the requirement for refined water system procedures.

As the twentieth century advanced, lift water system innovation kept on developing with the appearance of computerization and mechanization. Computerized control frameworks took into account exact administration of water stream, enhancing water system effectiveness and limiting water wastage. Remote detecting advancements and information examination further upgraded the checking and the executives of lift water system frameworks, introducing another period of brilliant horticulture.

In the 21st 100 years, the difficulties of environmental change, populace development, and expanding water shortage have moved the improvement of reasonable and eco-accommodating lift water system arrangements.

Advancements, for example, sunlight based fueled siphons and energy-proficient lift frameworks have acquired conspicuousness as the world looks to change to additional harmless to the ecosystem horticultural practices.

The joining of current sensor advancements, man-made reasoning, and information driven direction has additionally refined lift water system frameworks. These advances empower constant checking, examination, and versatile control, improving the effectiveness and manageability of water use in agribusiness. Accuracy farming practices, empowered by cutting edge sensors and computerized apparatus, add to streamlining water application in the field.

Regardless of the authentic importance and extraordinary effect of lift water system on horticulture, difficulties and debates persevere. The enormous scope redirection of water for water system raises worries about the consumption of normal water sources, environmental awkward nature, and clashes over water freedoms. Finding some kind of harmony between the requirement for water system and the maintainable administration of water assets stays a perplexing and progressing task for policymakers, specialists, and networks dependent on lift water system.

Taking everything into account, the advancement of lift water system frameworks across various societies and civilizations mirrors the unique transaction between human development, natural requirements, and the basic to guarantee food security. From antiquated manual gadgets and water wheels to steam-fueled motors and current computerized frameworks, the excursion of lift water system reflects the flexibility of agrarian practices to different geological and social settings. As we explore the intricacies of the 21st hundred years, the historical backdrop of lift water system gives important experiences into the continuous mission for proficient and reasonable water the board in farming.

The embroidery of mankind's set of experiences is woven with the strings of assorted societies and civic establishments, each contributing special tones, examples, and developments to the fantastic story of our aggregate process. As we navigate the hallways of time, the rich woven artwork unfurls, uncovering the bunch manners by which various societies and civic establishments have molded the course of humankind. From the support of old developments to the interconnected universe of the present, the development of social orders offers a significant look into the intricacies of human life.

In the support of progress, Mesopotamia, the Sumerians established the groundwork for metropolitan life and administration around 4500 BCE. The city-territories of Ur, Uruk, and Eridu prospered along the ripe banks of the Tigris and Euphrates waterways. The Sumerians, among the earliest composing societies, abandoned cuneiform engravings on earth tablets that uncover their complicated social design, legitimate codes, and strict convictions. The ziggurats, huge ventured structures committed to divinities, mirrored their engineering ability and profound tendencies.

Contemporaneously, in old Egypt along the banks of the Nile, an unmistakable human progress arose. The Pharaohs controlled a general public described by stupendous engineering, perplexing hieroglyphics, and a modern comprehension of heavenly peculiarities. The pyramids at Giza, the Sphinx, and the Valley of the Lords stand as persevering through demonstrations of their designing abilities and strict practices. The yearly flooding of the Nile, which brought rich sediment, was a help for horticulture and food.

Old China, with a civilization tracing all the way back to the Shang Tradition (1600-1046 BCE), unfurled along the Yellow Waterway. The Chinese fostered a complicated composing framework, farming strategies, and philosophical practices that keep on impacting this present reality. The idea of the Command of Paradise, administering the authenticity of rulers, and the lessons of Confucius and Laozi molded the moral and social texture of Chinese society.

In the Indian subcontinent, the Indus Valley Progress (3300-1300 BCE) flourished along the banks of the Indus Stream. Harappa and Mohenjo-daro were progressed metropolitan places with arranged roads, seepage frameworks, and amazing engineering. While the Indus script stays undeciphered, the development's monetary and social intricacy is apparent in its broad exchange organizations and particular specialty creation.

As we push ahead in time, the old style civilizations of Greece and Rome arise as reference points of scholarly and political idea. In old Greece, the city-territories of Athens and Sparta encapsulated unmistakable political frameworks. The origination of a majority rules system, Athens, cultivated scholarly request, delivering thinkers like Socrates, Plato, and Aristotle. The Olympic Games celebrated actual ability and fellowship, adding to the soul of Hellenic culture.

The Roman Republic and later the Roman Realm made a permanent imprint on Western human advancement. Roman regulation, designing wonders like water passages and streets, and the digestion of different societies inside its immense realm exhibit the regulatory and social ability of old Rome. The Pax Romana, a time of relative harmony, worked with exchange and social trades across the Mediterranean.

In South Asia, the Maurya and Gupta Realms (around 322 BCE - 550 CE) assumed key parts in forming the social and scholarly scene. The Mauryan head Ashoka, after a time of triumph, embraced Buddhism and proliferated its standards of peacefulness and resilience. The Gupta time frame is praised as the "Brilliant Age" of Indian human progress, seeing headways in science, arithmetic, and workmanship.

The Islamic Brilliant Age (eighth to fourteenth hundreds of years) remains as a demonstration of the thriving of information, workmanship, and development across the Islamic world. Researchers in Baghdad, Cairo, and Cordoba made huge commitments to fields like cosmology, medication, and theory.

The Place of Shrewdness in Baghdad filled in as a significant scholarly focus, deciphering Greek, Roman, and Indian texts into Arabic, saving and extending information.

In archaic Europe, the medieval framework, the ascent of governments, and the impact of the Catholic Church characterized the socio-political scene. The Gothic churches, like Notre-Lady in Paris, mirrored the otherworldly desires and structural ability of the age. The Magna Carta (1215) laid the foundation for sacred standards, while the Renaissance denoted a recovery of interest in traditional expressions and learning.

The Ming and Qing Traditions in China (1368-1912) saw times of social magnificence, financial flourishing, and international impact. The Ming Tradition, known for the development of the Incomparable Wall and sea investigations drove by Naval commander Zheng He, displayed China's mechanical and oceanic ability. The Qing Tradition, established by the Manchus, broadened Chinese domain and social impact.

In the Americas, native civilizations flourished well before European contact. The Maya, known for their high level schedule framework and amazing design, thrived in Mesoamerica. The Aztecs, with their capital Tenochtitlan, constructed a realm through military victory and created complex rural frameworks like chinampas, counterfeit islands utilized for development. The Inca Realm in South America, with its managerial wonder of the Machu Picchu, showed progressed designing and authoritative abilities.

The Time of Investigation in the fifteenth and sixteenth hundreds of years denoted a time of worldwide interconnectedness. European powers, impelled by sea progressions, wandered across seas looking for new shipping lanes and domains. Christopher Columbus' journey to the Americas in 1492 and resulting undertakings by travelers like Vasco da Gama and Ferdinand Magellan reshaped the international scene and started a course of social trade known as the Columbian Trade.

The Renaissance in Europe (fourteenth to seventeenth hundreds of years) saw a resurrection of scholarly and creative pursuits. Crafted by Leonardo da Vinci, Michelangelo, and Raphael exemplified a restored interest in humanism, science, and artistic expression. The creation of the print machine by Johannes Gutenberg altered the dispersal of information, democratizing admittance to data.

The Logical Upset (sixteenth to eighteenth hundreds of years) introduced another time of request and experimentation. Visionaries like Copernicus, Galileo, and Newton tested winning thoughts of the universe, propelling comprehension we might interpret cosmology and physical science. The Logical Strategy turned into a foundation of request, impacting disciplines past the inherent sciences.

The Illumination (seventeenth to eighteenth hundreds of years) accentuated reason, individual freedoms, and the quest for information. Thinkers like John Locke, Voltaire, and Jean-Jacques Rousseau supported for political and social changes. The Time of Edification laid the scholarly foundation for upheavals, including the American Insurgency (1775-1783) and the French Upset (1789-1799), which tried to lay out standards of a majority rules government and individual freedoms.

In the nineteenth hundred years, the Modern Unrest changed economies, social orders, and day to day existence. Steam motors, automated creation, and progressions in transportation changed assembling processes. The shift from agrarian economies to industrialized social orders reshaped work relations, urbanization designs, and worldwide exchange organizations.

Dominion and imperialism portrayed a large part of the nineteenth and mid twentieth hundreds of years, as European powers extended their realms across Africa, Asia, and the Americas. The scramble for assets, international predominance, and preacher energy molded the elements of supreme endeavors. The effect of colonization on native societies, economies, and political designs passed on an enduring inheritance that keeps on impacting worldwide relations.

The twentieth century saw extraordinary disturbances, including two Universal Conflicts and the Virus War, what isolated the world into philosophical coalitions drove by the US and the Soviet Association. The post-The Second Great War time saw the foundation of worldwide associations like the Unified Countries, pointed toward encouraging participation and forestalling worldwide struggles.

The Social equality Development in the US, drove by figures like Martin Luther Lord Jr., tested racial isolation and segregation, making ready for critical lawful and cultural changes. Additionally, the women's activist development upheld for orientation uniformity and ladies' privileges, igniting groundbreaking social movements.

In the last 50% of the twentieth hundred years, mechanical headways, especially in figuring and broadcast communications, introduced the Data Age. The improvement of the web and the multiplication of computerized innovations upset correspondence, business, and social connections on a worldwide scale.

The finish of the Virus Battle in 1991 denoted the breakdown of the Soviet Association and the rise of a unipolar world request with the US as the prevailing superpower. In any case, the 21st century has seen a change in international elements, with the ascent of arising powers, monetary relationship, and transnational difficulties, for example, environmental change, psychological oppression, and pandemics.

As we explore the intricacies of the contemporary world, various societies and civic establishments proceed to impact and meet with each other. Globalization, worked with by propels in transportation and correspondence, has made an interconnected reality where thoughts, products, and societies cross lines no sweat.

Social variety, once bound to nearby settings, is currently a worldwide peculiarity. The trading of thoughts, relocation designs, and the mixing of social impacts have led to a multicultural embroidery that rises above customary limits.

Chapter 3

Engineering Wonders

Designing has for quite some time been a main thrust behind human advancement, pushing the limits of what is conceivable and molding the world we live in. Since forever ago, engineers have considered, planned, and constructed designs and innovations that stand as demonstration of human resourcefulness and desire. These designing marvels, dispersed across the globe, act as achievements in the development of our human progress, mirroring the headways in science, innovation, and craftsmanship.

One such wonder is the Incomparable Mass of China, a goliath demonstration of old designing ability. Extending north of 13,000 miles across rough landscape, this impressive design was worked more than a few traditions to guard against intrusions. The Incomparable Wall epitomizes the sheer scale and assurance of human undertaking, utilizing a blend of regular materials and vital plan to make an imposing safeguard framework.

Pushing ahead in time, the Pyramids of Giza in Egypt arise as another designing victory. Developed quite a while back, these monstrous designs have endured everyday hardship, filling in as burial chambers for pharaohs. The accuracy with which the old Egyptians adjusted the pyramids to cosmic peculiarities is a demonstration of their refined comprehension of science and designing.

As civic establishments advanced, so did designing accomplishments. The Roman Colosseum, a notorious amphitheater, remains as an image of design and designing development. Finished in Promotion 80, this gigantic field could oblige up to 80,000 onlookers, facilitating combatant challenges and public exhibitions. The brilliant utilization of curves, vaults, and cement permitted the Romans to make amazing designs that would move modelers long into the future.

The Medieval times introduced an alternate period of designing wonders, with houses of God ascending to overwhelm European horizons. Chartres Church building in France, worked in the thirteenth 100 years, is a great representation of Gothic design. Its taking off towers, mind boggling stained glass windows, and flying supports feature the combination of imaginative magnificence and underlying development. These basilicas were spots of love as well as demonstrations of the abilities of archaic designing.

The Renaissance achieved a recovery of traditional information and a flood in mechanical development. Leonardo da Vinci, a polymath of the period, abandoned a store of designing portrayals and plans. His flying machines, pressure driven frameworks, and physical investigations laid the basis for future architects and designers. The Renaissance time additionally saw the development of wonders like the Inclining Pinnacle of Pisa, which, regardless of its primary blemish, stays remaining as a getting through image of engineering dauntlessness.

The Modern Upheaval denoted a defining moment in designing history. Steam motors, rail lines, and plants changed social orders, prompting the ascent of urban communities and the automation of work. The Gem Castle in London, worked for the Incomparable Presentation of 1851, exhibited the conceivable outcomes of iron and glass in design. This straightforward and roomy construction was a wonder of now is the right time, addressing the combination of industry and plan.

In the twentieth hundred years, designing arrived at new levels with the approach of high rises. The Eiffel Pinnacle, finished in 1889, was an image of French development as well as a spearheading illustration of iron development. The race for taller structures escalated in the next many years, finishing in the development of the Realm State Working in 1931. This Workmanship Deco show-stopper remained as the world's tallest structure for almost forty years, displaying the headways in primary designing and materials.

The mid-twentieth century achieved an upheaval in transportation with the improvement of the interstate expressway framework. The development of enormous extensions, for example, the Brilliant Door Scaffold in San Francisco, became images of network and progress. These engineered overpasses, with their general ranges and rich plans, pushed the limits of what was viewed as conceivable in structural designing.

The Space Age introduced another period of investigation and mechanical accomplishment. The Apollo 11 mission, which effectively landed people on the moon in 1969, was a demonstration of the cooperative endeavors of specialists, researchers, and space explorers. The Saturn V rocket, a transcending behemoth of designing, pushed the space apparatus into space, stamping perhaps of humankind's most prominent accomplishment.

As the 21st century unfurls, designing keeps on molding the world in exceptional ways. The Burj Khalifa in Dubai, finished in 2010, remains as the tallest structure on earth, an image of building desire and mechanical ability. Its smooth plan and inventive utilization of materials mirror the potential outcomes of current designing.

Progressions in materials science have additionally upset framework projects. The Millau Viaduct in France, finished in 2004, is the world's tallest extension, traversing a stunning crevasse. The utilization of high-strength, lightweight materials permitted specialists to make a design that appears to challenge gravity, exhibiting the marriage of style and usefulness.

The field of advanced mechanics and man-made consciousness has opened up new boondocks in designing. Robots are currently utilized in different enterprises, from assembling to medical care. The improvement of independent vehicles, fueled by artificial intelligence calculations, is ready to alter transportation and reshape metropolitan scenes. These advancements address the front line of designing, pushing the limits of what is conceivable.

In the domain of energy, engineers are handling the test of feasible power sources. The ascent of sun based and wind energy advancements is changing the manner in which we create and consume power. Enormous sunlight based homesteads and wind turbines, with their exquisite yet strong plans, are becoming pervasive images of a shift towards a more economical future.

Designing miracles are not restricted to Earth alone. The Worldwide Space Station (ISS), a cooperative exertion including different nations, circles the Earth as a demonstration of human participation and designing greatness. This livable satellite is a wonder of room designing, giving a stage to logical examination and global cooperation.

The journey for clean energy has additionally prompted the improvement of combination reactors. ITER, a worldwide venture situated in France, means to show the possibility of atomic combination as a manageable and bountiful wellspring of force.

The designing difficulties engaged with binding and controlling high-temperature plasma are massive, however achievement could reform the world's energy scene.

In the domain of data innovation, the improvement of quantum PCs is ready to upset processing power. These machines, utilizing the standards of quantum mechanics, can possibly tackle complex issues that are at present past the span of traditional PCs. The competition to construct viable quantum PCs is a demonstration of the persistent quest for development in the field of designing.

The designing wonders of today are not just about great designs and amazing accomplishments. Scaling down and microengineering have brought about strong gadgets that fit in the center of our hands. Cell phones, with their complex equipment and programming, are a demonstration of the union of numerous designing disciplines. The fast speed of mechanical headway in the gadgets business keeps on rethinking the conceivable outcomes of what can be accomplished on a limited scale.

Biomedical designing is another boondocks where advancement is quickly evolving lives. Progresses in clinical imaging, prosthetics, and hereditary designing are changing medical services and pushing the limits of what was once imagined. 3D printing innovation, specifically, has opened up additional opportunities in the production of modified inserts and organs, exhibiting the capability of designing to improve and expand human existence.

Despite environmental change and natural difficulties, engineers are investigating inventive arrangements. Green engineering, portrayed by supportable and energy-productive plan, is turning out to be progressively pervasive. Structures with green rooftops, sun powered chargers, and high level protection frameworks are tastefully satisfying as well as add to the more extensive objective of diminishing the ecological effect of human exercises.

The designing miracles representing things to come might reach out past our planet. The idea of room colonization is getting forward movement, with designers and researchers imagining human settlements on Mars and then some. The difficulties of making reasonable environments in space, with shut circle life emotionally supportive networks and high level advanced mechanics, present extraordinary designing difficulties. The fantasy about turning into a multi-planetary animal categories is pushing the limits of what designing can accomplish.

All in all, the historical backdrop of designing is an account of human desire, development, and persistence. From antiquated ponders like the Incomparable Mass of China to current wonders like the Burj Khalifa, engineers have formed the world we possess. The interdisciplinary idea of designing, drawing from fields as different as physical science, materials science, science, and software engineering, keeps on driving advancement and open up new outskirts.

As we stand on the cusp of another time, with arising advances like quantum figuring, space investigation, and manageable energy, the job of designing in molding what's to come is more basic than any time in recent memory. The difficulties that lie ahead, from addressing environmental change to investigating the universe, will require the aggregate aptitude and innovativeness of architects all over the planet.

3.1 In-depth exploration of the engineering principles behind lift irrigation

Lift water system remains as a demonstration of the resourcefulness of designing in enhancing water use for farming, especially in regions with testing geology. This strategy includes raising water from a lower rise to a higher one, permitting it to course through channels and arrive at fields that sounds distant. The designing standards behind lift water system are complex, enveloping parts of liquid mechanics, foundational layout, and energy the executives.

At its center, lift water system tends to the principal challenge of water shortage in areas where customary strategies for water system miss the mark. In numerous rural scenes, the accessibility of water is unevenly conveyed, and the territory might present deterrents to the normal progression of water. Lift water system frameworks give an answer by utilizing designing standards to defeat these difficulties and guarantee a dependable water supply for crops.

The principal basic part of lift water system lies in the comprehension of liquid mechanics. Water, as a liquid, follows the standards of hydrodynamics, represented by variables like tension, stream, and height. Engineers planning lift water system frameworks should compute the head, which addresses the energy accessible to lift the water. This head can be gotten from the level distinction between the water source and the most elevated point in the water system organization.

Siphons assume a focal part in lift water system frameworks, filling in as the mechanical gadgets that raise water to the ideal level. The determination of a suitable siphon relies upon different variables, including the expected stream rate, the upward lift, and the kind of force source accessible. Radial siphons are generally utilized in lift water system because of their proficiency and capacity to deal with shifting stream rates.

The energy expected to work the siphons is a basic thought in the general effectiveness of lift water system frameworks. Engineers should cautiously evaluate the power source choices, which might incorporate electric engines, diesel motors, or inexhaustible sources, for example, sunlight based or wind energy. The decision of force source relies upon factors like accessibility, cost, and ecological contemplations. As of late, there has been a developing accentuation on coordinating feasible and sustainable power arrangements into lift water system frameworks to diminish their natural effect.

Underlying designing is one more pivotal part of lift water system frameworks. The organization of lines, channels, and supplies should be intended to endure the powers applied by the streaming water and the outer components. The arrangement of channels and the position of lines are basic to improving the stream and limiting energy misfortunes. Engineers use standards of liquid elements to configuration channels that limit frictional misfortunes and guarantee productive water transport.

One of the critical difficulties in lift water system is moderating the impacts of height changes on water pressure. As water is lifted to higher heights, the strain diminishes, possibly prompting issues like cavitation in siphons. Engineers utilize different techniques, including the utilization of various siphoning stages and the consolidation of tension help components, to keep up with ideal strain levels and forestall harm to the framework.

Supply configuration is a basic part of lift water system framework. The supply fills in as a storeroom, permitting water to be gathered during times of overflow and delivered depending on the situation during droughts. Engineers should cautiously compute the limit of the repository, considering variables, for example, the water interest of yields, the span of the dry season, and the inflow rate from water sources.

In bumpy or sloping landscape, the geology acquaints extra difficulties with lift water system. Engineers should plan frameworks that explore steep inclines, frequently requiring the utilization of particular designs like siphons and reservoir conduits. These designs should be painstakingly designed to endure the powers applied by the streaming water and the land states of the territory.

Computerization and control frameworks assume an essential part in upgrading the activity of lift water system frameworks. Engineers consolidate sensors, actuators, and control calculations to screen water levels, manage siphon activity, and guarantee proficient water dissemination. Propels in innovation, including the Web of Things

(IoT) and information examination, empower constant observing and direction, upgrading the general exhibition of lift water system frameworks.

The social and monetary parts of lift water system can't be ignored. Engineers should consider the requirements and inclinations of the networks that depend on these frameworks for their jobs. Local area commitment is fundamental in the preparation and execution stages, guaranteeing that the designing arrangements line up with the neighborhood setting and add to feasible farming practices.

Lift water system frameworks additionally have natural ramifications, and architects should endeavor to limit their biological impression. Endeavors to lessen energy utilization, use environmentally friendly power sources, and carry out water-saving innovations add to the general maintainability of these frameworks.

Also, ecological effect evaluations are essential to understanding and moderating any possible antagonistic consequences for environments and regular water sources.

Contextual analyses of fruitful lift water system projects give important bits of knowledge into the pragmatic utilization of designing standards. The Sardar Sarovar Dam in India, for instance, is a stupendous lift water system project that tackles the waters of the Narmada Stream to flood immense horticultural regions in the western provinces of Gujarat and Rajasthan. The dam's complicated organization of channels and pipelines uses the standards of liquid elements and underlying designing to convey water to ranchers across the locale.

In China, the Dujiangyan water system framework remains as a demonstration of old designing insight. Worked quite a long time back, this gravity-based water system framework redirects water from the Minjiang Stream to the Chengdu Plain, giving a dependable water supply to farming. The outcome of Dujiangyan lies in its shrewd plan, which takes advantage of the regular geography to disseminate water without the requirement for siphons.

The effect of lift water system reaches out past horticulture to impact more extensive financial turn of events. Dependable admittance to water improves crop yields, upholds food security, and gives open doors to financial development in rustic networks. Engineers engaged with lift water system projects assume a vital part in tending to the perplexing exchange of specialized, social, and ecological variables to make maintainable arrangements.

Looking forward, the fate of lift water system is entwined with progressing headways in designing and innovation. The mix of shrewd advancements, accuracy farming strategies, and the utilization of computerized reasoning can additionally improve the productivity of water use in horticulture. Advancements in materials science might prompt the improvement of more strong and manageable foundation for lift water system frameworks.

As worldwide water provokes escalate because of environmental change and populace development, the job of lift water system turns out to be progressively basic. Designers will keep on being at the very front of creating arrangements that guarantee water security for rural networks. The interdisciplinary idea of lift water system designing, joining aptitude in hydrodynamics, underlying model, energy the board, and financial contemplations, highlights the intricacy of tending to water shortage through creative designing arrangements.

All in all, lift water system remains as an exceptional designing accomplishment that tends to the complex difficulties of water conveyance in horticulture. From the standards of liquid mechanics and siphon innovation to the complexities of foundational layout and energy the executives, lift water system frameworks exemplify the cooperative endeavors of specialists to conquer geological and geographical requirements.

As these frameworks develop with mechanical headways and a developing consciousness of natural supportability, they keep on assuming a crucial part in supporting rural turn of events and upgrading water versatility in different scenes all over the planet.

3.2 Profiles of notable lift irrigation projects around the world

Lift water system projects play had an essential impact in changing bone-dry and testing scenes into prolific rural districts, guaranteeing water access for harvests and networks. Across the globe, different prominent lift water system projects stand as designing wonders, exhibiting imaginative answers for address water shortage and streamline rural efficiency.

One commendable task is the Sardar Sarovar Dam in India, a huge pressure driven designing undertaking that outfits the waters of the Narmada Waterway. This multi-purpose task, started during the 1960s, fills in as a help for the western territories of Gujarat and Rajasthan. The Sardar Sarovar Dam, perhaps of the biggest dam on the planet, integrates lift water system as a critical part to disseminate water across immense farming regions.

The dam, with a level of 138 meters, makes an enormous repository that can save to 9.5 billion cubic meters of water. This repository goes about as a support, putting away water during the storm season and delivering it decisively over time. The lift water system network associated with the dam contains a complex arrangement of trenches and pipelines that transport water to remote fields, defeating the difficulties presented by the district's undulating geography.

Diffusive siphons, decisively positioned along the organization, assume a critical part in lifting water to the expected heights. These siphons are controlled by electric engines, taking advantage of the lattice for a ceaseless and solid energy source. The size of the Sardar Sarovar Dam lift water system project is monstrous, covering huge number of kilometers of channels and helping a great many ranchers in the locale.

The outcome of the Sardar Sarovar Dam project lies in its designing complexities as well as in the thorough way to deal with water the executives. The undertaking coordinates hydropower age, water supply for homegrown and modern use, and water system for agribusiness. This complex use of water assets exhibits the capability of lift water system ventures to address different requirements and add to territorial turn of events.

In China, the Dujiangyan water system framework remains as a demonstration of old designing insight and the getting through effect of lift water system. Built a while back, during the Qin Line, Dujiangyan stays a utilitarian and fundamental piece of the Chengdu Plain's horticultural scene. The framework redirects water from the Minjiang Stream without the requirement for a dam, depending on normal geology and cunning designing.

The critical element of Dujiangyan is the Yuzui (Fish Mouth) Redirection Channel, which bifurcates the waterway into an internal and external branch. The internal branch supplies water straightforwardly to the Chengdu Plain, while the external branch forestalls flooding and sedimentation. This inventive plan takes into consideration the controlled arrival of water without the requirement for a siphon, making Dujiangyan a gravity-based lift water system framework.

The progress of Dujiangyan lies in its supportable and low-upkeep plan, a demonstration of the old's comprehension designers might interpret water powered standards. The framework has endured everyday hardship, giving dependable water supply to the area and adding to the success of Chengdu, one of China's significant urban communities.

Moving to the African landmass, the Gezira Plan in Sudan is another critical lift water system project with a rich history. Started during the English frontier time in the mid twentieth hundred years, the Gezira Plan expected to change the bone-dry grounds along the Blue Nile into a useful rural locale. The task used the waters of the Blue Nile, utilizing an organization of trenches and siphons to lift water to higher heights.

The Gezira Plan covers a broad area of in excess of 8,000 square kilometers and has been a pivotal supporter of Sudan's rural result. Cotton, wheat, and different yields flourish in this watered scene, supporting the occupations of thousands of ranchers. The task addresses a cooperative exertion between specialists, agronomists, and policymakers to address the water difficulties in a district described by an articulated dry season.

In the US, the Focal Valley Task (CVP) in California represents the utilization of lift water system in a cutting edge rural setting. Started during the 1930s, the CVP meant to give water to the dry Focal Valley, perhaps of the most useful farming district in the country. The undertaking includes the development of dams, channels, and siphoning stations to lift water from the Sacramento-San Joaquin Stream Delta to the farming grounds of the Focal Valley.

The CVP integrates both gravity-based and lift water system parts, using the regular incline of the land where conceivable and utilizing siphons where rise contrasts present difficulties. The framework upholds a different scope of yields, including natural products, vegetables, and nuts, contributing essentially to California's rural result. In any case, the undertaking has likewise been a subject of discussion because of its effect on biological systems, water designation issues, and natural supportability.

In Israel, where water shortage is a consistent test, creative ways to deal with lift water system have been created. The Netafim Trickle Water system Framework, albeit not a conventional lift water system project, represents accuracy water conveyance to farming fields. Netafim's innovation includes the utilization of trickle lines to convey water straightforwardly to the foundations of plants, limiting wastage and enhancing water use proficiency.

While not a lift water system framework in the conventional sense, the Netafim approach shows the way that designing development can address water shortage issues. By definitively controlling the conveyance of water to crops, this framework saves assets and permits horticulture to flourish in areas with restricted water accessibility.

The progress of lift water system projects not entirely settled by their designing complexities; it is additionally dependent upon successful administration and local area commitment. The Bhakra Nangal Dam in India is a multi-reason project that incorporates lift water system parts and epitomizes the significance of complete preparation and the executives.

Based on the Sutlej Waterway, the Bhakra Nangal Dam is quite possibly of the biggest dam in India, making a repository that traverses the territories of Punjab, Haryana, and Rajasthan. The dam consolidates lift water system to supply water to horticultural terrains in these states, supporting a scope of harvests, including rice and wheat. The venture's prosperity is credited not exclusively to the designing ability yet additionally to the effective water the board rehearses and the helpful endeavors of the states in question.

In Brazil, the São Francisco Stream Mix Undertaking addresses a huge lift water system drive pointed toward changing the semi-dry grounds of northeastern Brazil. The task includes the development of trenches, siphoning stations, and repositories to redirect water from the São Francisco Stream to bone-dry districts in the provinces of Pernambuco, Ceará, Paraíba, and Rio Grande do Norte.

The reconciliation of lift water system parts in the São Francisco Stream Venture is fundamental for passing water on to higher heights and guaranteeing inclusion of broad rural regions. The undertaking has confronted ecological and social difficulties, remembering worries about the effect for neighborhood environments and relocation of networks. In any case, it highlights the aspiration of designing answers for address water shortage and encourage horticultural improvement in testing conditions.

In the domain of lift water system, mechanical headways keep on forming the scene of conceivable outcomes. The utilization of sunlight based controlled siphons is getting forward momentum as a practical and decentralized way to deal with lifting water. Sun based lift water system frameworks, especially in off-network and distant regions, exhibit the possibility to diminish dependence on regular energy sources and improve the versatility of rural networks.

Furthermore, the reconciliation of brilliant advancements and information examination is changing how lift water system frameworks are checked and made due. Constant sensors, mechanized controls, and prescient calculations empower specialists to improve water appropriation, distinguish irregularities, and answer proactively to evolving conditions. These progressions add to the effectiveness, manageability, and versatility of lift water system projects despite dynamic natural elements.

As lift water system projects keep on developing, interdisciplinary coordinated effort turns out to be progressively pivotal. Engineers, hydrologists, agronomists, and policymakers should cooperate to plan and execute arrangements that offset horticultural requirements with natural supportability. The examples gained from the profiles of outstanding lift water system projects all over the planet highlight the significance of setting explicit methodologies, local area contribution, and an all encompassing comprehension of the intricate interaction between water, farming, and cultural necessities.

All in all, lift water system projects address a union of designing skill, natural contemplations, and financial objectives. From old frameworks like Dujiangyan in China to current undertakings like the Sardar Sarovar Dam in India, these tasks embody the versatility of human resourcefulness in conquering water difficulties. As the world countenances developing water shortage and changing environment designs, lift water system stays a basic apparatus in the worldwide work to guarantee food security, cultivate supportable horticulture, and upgrade the versatility of networks in different scenes.

3.3 Examination of technological advancements in lift irrigation systems

Mechanical headways in lift water system frameworks have been instrumental in upgrading effectiveness, manageability, and versatility. As worldwide water difficulties heighten and the interest for horticultural efficiency rises, specialists and analysts have utilized state of the art innovations to advance the plan, activity, and the executives of lift water system projects. This assessment digs into key mechanical headways that are forming the scene of lift water system frameworks around the world.

One of the critical steps in lift water system innovation is the combination of accuracy horticulture procedures. Accuracy horticulture use data innovation, sensors, and information examination to advance the utilization of water, manures, and different assets in agribusiness. When applied to lift water system, accuracy horticulture empowers ranchers and water chiefs to fit water conveyance to the particular requirements of yields, guaranteeing ideal development while limiting wastage.

The organization of soil dampness sensors, weather conditions stations, and remote detecting advancements considers constant checking of soil conditions and yield water necessities. This information driven approach empowers exact independent direction in regards to when and how much water to convey, forestalling over-water system and waterlogging. High level calculations investigate the gathered information, giving bits of knowledge that assist ranchers and water chiefs with pursuing informed decisions to augment crop yields and water use proficiency.

Mechanization assumes a crucial part in improving the functional productivity of lift water system frameworks. Computerized control frameworks, frequently coordinated with sensors and actuators, empower remote checking and the board of the whole water system organization.

This not just decreases the requirement for on location intercessions yet in addition considers quick reactions to evolving conditions, guaranteeing that water is conveyed where and when it is generally required.

Savvy siphons outfitted with variable recurrence drives (VFDs) address one more mechanical jump in lift water system. VFDs take into account the exact control of siphon speed and power utilization, empowering versatile reactions to fluctuating water interest. This innovation improves energy effectiveness, as siphons can work at ideal paces in view of ongoing necessities. Furthermore, brilliant siphons can be incorporated into computerized control frameworks, working with consistent coordination inside the general lift water system foundation.

Sustainable power sources are progressively being incorporated into lift water system frameworks, offering manageable options in contrast to traditional power supplies. Sun oriented controlled siphons, specifically, have acquired conspicuousness in off-lattice and far off regions where admittance to power is restricted. These siphons saddle sunlight based energy to drive the lifting instrument, diminishing dependence on ordinary energy sources and bringing down functional expenses. As sun based innovation keeps on propelling, the proficiency and moderateness of sun oriented fueled lift water system frameworks are probably going to work on further.

Headways in materials science have added to the advancement of additional sturdy and erosion safe parts for lift water system framework. The lines, siphons, and valves utilized in lift water system frameworks are presented to a scope of natural circumstances, and the selection of materials is basic to guaranteeing life span and negligible upkeep. Developments in materials, like the utilization of high-thickness polyethylene (HDPE) for pipes and progressed composites for siphons, upgrade the versatility and life expectancy of lift water system foundation.

Web of Things (IoT) innovations are upsetting how lift water system frameworks are observed and made due. IoT gadgets, including sensors and actuators, can be conveyed all through the water system organization to gather continuous information on water stream, pressure, and other basic boundaries. This information is sent to a focal control framework, considering far reaching observing and empowering opportune changes in accordance with enhance water dissemination. The network given by IoT innovations upgrades the general knowledge of lift water system frameworks, preparing for additional responsive and versatile activities.

Remote detecting advancements, like satellite symbolism and automated elevated vehicles (UAVs), offer significant experiences for lift water system the board. These advances give high-goal symbolism of agrarian fields, taking into consideration the appraisal of harvest wellbeing, water pressure, and generally speaking vegetation conditions. By incorporating remote detecting information into the dynamic cycle, water directors can acquire a thorough comprehension of the water system scene and pursue informed decisions to address explicit difficulties.

The reception of versatile applications and cloud-based stages has changed the manner in which ranchers and water administrators associate with lift water system frameworks. Portable applications can give continuous updates on water accessibility, conveyance timetables, and yield conditions. Cloud-based stages offer concentrated information capacity and examination, working with coordinated effort and data dividing between partners. These computerized apparatuses engage clients to settle on informed choices, improve asset portion, and upgrade the general exhibition of lift water system frameworks.

Man-made brainpower (computer based intelligence) and AI calculations are progressively being applied to lift water system frameworks, offering progressed examination and prescient abilities. Simulated intelligence calculations can dissect immense datasets, including authentic atmospheric conditions, crop execution, and water use, to distinguish patterns and examples. This prescient ability empowers water supervisors to expect future water system needs, improve planning, and proactively address likely difficulties. AI calculations, through ceaseless gaining from constant information, add to continuous enhancements in framework proficiency.

The idea of blockchain innovation is arising as an expected answer for upgrade straightforwardness and responsibility in water the executives, including lift water system projects. Blockchain, a decentralized and secure record framework, can be applied to record and confirm water exchanges, guaranteeing discernibility and forestalling unapproved modifications. This innovation holds guarantee in making a straightforward and alter safe record of water distributions, moves, and use, encouraging trust among partners in the lift water system environment.

Drones, or automated elevated vehicles (UAVs), have tracked down applications in observing and overseeing lift water system frameworks. Outfitted with high-goal cameras and sensors, robots can catch definite symbolism of fields, water system foundation, and water circulation designs. This elevated viewpoint gives significant experiences into the state of harvests, the productivity of water system rehearses, and the honesty of framework. Drones offer a savvy and proficient method for looking over enormous regions, working with opportune intercessions and enhancing asset portion.

Environment displaying and prescient examination are becoming basic parts of cutting edge lift water system frameworks. By utilizing verifiable environment information, meteorological figures, and high level demonstrating methods, specialists and water chiefs can expect changes in weather conditions and their effect on water accessibility. This premonition considers proactive acclimations to water system plans, guaranteeing that water assets are overseen actually despite advancing environment conditions.

The rise of 5G innovation holds guarantee for changing correspondence and information move inside lift water system frameworks. The rapid, low-idleness availability presented by 5G organizations empowers constant correspondence between sensors, control frameworks, and different parts of the water system foundation.

This availability improves the responsiveness of robotized frameworks, works with prompt information move, and supports the joining of arising innovations, like expanded reality and computer generated reality, for upgraded observing and upkeep.

Expanded reality (AR) and augmented reality (VR) innovations offer novel ways to deal with picturing and overseeing lift water system framework. AR applications can overlay advanced data onto the actual climate, giving ongoing bits of knowledge into the situation with siphons, lines, and valves. VR reenactments permit designers and administrators to essentially explore through the water system organization, distinguish expected issues, and plan support exercises. These vivid advances add to advanced situational mindfulness and more successful direction.

The idea of computerized twins, virtual imitations of actual frameworks, is getting some forward movement in the domain of lift water system. Advanced twins empower continuous demonstrating and reproduction of the whole water system foundation, consolidating information from sensors and different sources. This virtual portrayal considers persistent checking, investigation, and enhancement of framework execution. By reflecting the actual reality, computerized twins give a useful asset to figuring out complex collaborations inside lift water system frameworks.

The mix of organic arrangements, like biotechnology and hereditary designing, is an arising outskirts in upgrading crop strength with regards to lift water system. Researchers are investigating the advancement of yields with further developed dry season opposition, water use effectiveness, and resistance to saltiness. These hereditarily altered crops can possibly flourish in water-scant conditions, decreasing the general water interest in lift water system projects and adding to feasible agribusiness.

The idea of nanotechnology is being investigated for its expected application in water filtration and purging inside lift water system frameworks. Nanomaterials, with their extraordinary properties, can be utilized to make progressed channels that actually eliminate impurities from water system water. Nanoscale sensors can likewise be coordinated into water system foundation to screen water quality and distinguish possible issues. The utilization of nanotechnology adds to the productive and feasible administration of water assets in lift water system.

As lift water system frameworks keep on developing, the coordination of these advances offers uncommon open doors for further developing execution, manageability, and strength. Notwithstanding, it is fundamental to recognize the difficulties related with embracing trend setting innovations, including issues connected with moderateness, availability, and the requirement for talented faculty. Moreover, moral contemplations and the possible ecological effects of specific innovations should be painstakingly assessed.

All in all, mechanical headways are assuming an extraordinary part in the domain of lift water system frameworks. From accuracy agribusiness and mechanization to computerized reasoning and arising advancements like blockchain and nanotechnology, these developments are reshaping how water is made due, conveyed, and used in horticulture. The continuous reconciliation of cutting edge innovations holds the commitment of making more effective, feasible, and versatile lift water system frameworks that can address the developing difficulties of water shortage, environmental change, and food security on a worldwide scale.

Lift water system frameworks address a basic feature of horticultural foundation intended to address water shortage and work with productive water circulation to farmlands. These frameworks assume an essential part in districts where the normal progression of water is deficient to satisfy the needs of farming, particularly in regions with testing geography. An itemized investigation of lift water system frameworks envelops their verifiable development, designing standards, functional parts, and the financial effect they employ in supporting horticultural exercises.

The underlying foundations of lift water system frameworks can be followed back to antiquated civic establishments that perceived the significance of water the executives in farming. Early social orders, for example, those in Mesopotamia and the Indus Valley, created simple strategies to lift water from waterways and streams to flood fields. Straightforward gadgets like the shaduf in antiquated Egypt and the Persian wheel in Persia were early appearances of lift water system innovation, exhibiting the creativity of old designers in outfitting water for horticulture.

Throughout the long term, the complexity of lift water system frameworks developed. In bygone eras, the Islamic Brilliant Age saw headways in water designing, with the improvement of more productive water-lifting gadgets like the noria, a kind of water wheel. These developments were necessary to supporting farming in bone-dry and semi-dry locales, mirroring the nearby association between water the executives and horticultural efficiency.

The cutting edge time introduced another rush of designing developments in lift water system. The coming of the Modern Transformation in the eighteenth century denoted a defining moment, achieving motorization and mechanical headways that would reform farming. Steam motors and siphons fueled by steam became instrumental in lifting water to higher rises, growing the range of water system to already distant regions.

The designing standards behind lift water system frameworks rotate around the essential laws of liquid mechanics, power through pressure, and energy move. The center test is to lift water from a lower height to a higher one, defeating the power of gravity. Engineers utilize various techniques and advancements to accomplish this, each custom-made to the particular requirements and states of the scene.

Divergent siphons are a typical part of lift water system frameworks, utilizing the standards of liquid elements. These siphons utilize rotational dynamic energy to grant speed to water, making a stream that can lift water to higher heights. The determination of a proper siphon is dependent upon elements, for example, the expected stream rate, the lift level, and the power source accessible. Outward siphons are leaned toward for their productivity, unwavering quality, and capacity to deal with fluctuating stream rates.

In districts where power is promptly accessible, electric engines power the siphons. Diesel motors are an elective when electrical framework is restricted or temperamental. Also, the combination of sustainable power sources, for example, sunlight based and wind power, is building up some decent momentum, adding to the maintainability of lift water system frameworks. The decision of force source is a basic thought, affecting the general effectiveness, functional expense, and ecological effect of the framework.

The idea of head, addressing the energy accessible to lift water, is fundamental to the plan of lift water system frameworks. Head is determined in view of the level distinction between the water source and the most noteworthy point in the water system organization. The higher the head, the more energy is accessible for lifting water. Designs cautiously assess the geography of the territory to decide the ideal areas for water sources, supplies, and dissemination focuses, planning to augment the accessible head.

Repositories are vital parts of lift water system foundation, filling in as storage spaces for water. These repositories gather water during times of overflow, like the blustery season, and delivery it depending on the situation during droughts. The limit of the repository is a basic thought, dependent upon factors like the water interest of harvests, the span of the dry season, and the inflow rate from water sources. Engineers endeavor to figure out some kind of harmony that guarantees a steady and dependable water supply all through the horticultural cycle.

The plan and development of trenches and pipelines are fundamental parts of lift water system projects. Engineers should cautiously design the arrangement of channels, taking into account factors, for example, slant, height changes, and frictional misfortunes. The utilization of materials impervious to consumption and wear is significant for the life span and strength of the water system framework. The arrangement of trenches and the distance across of lines are upgraded to limit frictional misfortunes and guarantee proficient water transport.

In areas with testing geology, lift water system frameworks might include the utilization of particular designs like reservoir conduits, siphons, and passages. These designing arrangements are formulated to explore steep inclines and convey water across impediments. Reservoir conduits, for example, are raised channels that convey water over valleys and lopsided landscape. The plan and development of such designs require a nuanced comprehension of topographical circumstances and pressure driven standards.

Computerization and control frameworks are progressively being incorporated into lift water system tasks to upgrade functional productivity. Sensors, actuators, and control calculations empower constant checking of water levels, siphon activity, and by and large framework execution. Computerized control frameworks work with exact guideline of water stream and conveyance, limiting wastage and advancing asset use. The coordination of these advances engages water supervisors to answer quickly to evolving conditions, adding to the flexibility of lift water system frameworks.

Local area commitment is a significant element of lift water system projects, perceiving the social and monetary effect on neighborhood populaces. Ranchers, as essential partners, assume a functioning part in the preparation, execution, and the board of lift water system frameworks. Their insight into neighborhood agrarian practices, water needs, and land use designs is significant for planning frameworks that line up with the local area's prerequisites and needs.

The financial effect of lift water system frameworks stretches out past agribusiness to impact more extensive country improvement. Solid admittance to water upgrades crop yields, upholds food security, and gives open doors to monetary development in country networks. Lift water system projects add to neediness lightening by cultivating horticultural efficiency, setting out work open doors, and upgrading the general personal satisfaction in country regions.

As lift water system frameworks keep on developing, the reconciliation of cutting edge innovations is molding the future direction of these undertakings. Accuracy horticulture methods, empowered by sensors, information examination, and mechanization, are turning out to be progressively common. The utilization of robots for elevated studies and observing gives a 10,000 foot perspective of fields and foundation, helping with direction and support arranging.

Brilliant water system frameworks, utilizing Web of Things (IoT) innovations, empower continuous observing and control of water conveyance. These frameworks change water system plans in light of weather patterns, soil dampness levels, and harvest prerequisites, enhancing water use productivity. The network given by IoT advancements upgrades the general knowledge of lift water system frameworks, cultivating more educated direction and asset the executives.

Sun based fueled siphons address a feasible option for controlling lift water system frameworks, especially in off-network and far off regions. These siphons outfit energy from the sun to lift water, diminishing reliance on regular energy sources and bringing down functional expenses. The progression of sun based innovation and enhancements in energy capacity arrangements add to the feasibility and reasonableness of sun oriented controlled lift water system frameworks.

Progresses in materials science add to the improvement of additional sturdy and strong parts for lift water system framework. Consumption safe materials, built up composites, and high level coatings improve the life expectancy of lines, siphons, and valves. The utilization of imaginative materials mitigates support prerequisites and guarantees the drawn out unwavering quality of lift water system frameworks.

The joining of computerized reasoning (man-made intelligence) and AI calculations offers prescient abilities for lift water system the executives. These calculations examine authentic information, weather conditions, and harvest execution to anticipate future water system needs. AI models ceaselessly gain from continuous information, empowering versatile reactions and progressing upgrades in framework productivity.

Blockchain innovation, with its decentralized and secure record framework, holds potential for upgrading straightforwardness and responsibility in water the board. In lift water system projects, blockchain can be applied to record and check water exchanges, guaranteeing recognizability and forestalling unapproved changes. This innovation cultivates trust among partners and adds to more evenhanded water circulation.

The idea of computerized twins, virtual reproductions of actual frameworks, is arising as an incredible asset for lift water system the executives. Advanced twins empower ongoing displaying and reproduction of the whole water system foundation, giving experiences into framework conduct and execution. This virtual portrayal works with constant observing, investigation, and improvement of lift water system frameworks.

Chapter 4

Benefits and Challenges

The reconciliation of innovation into different parts of our lives has achieved huge advantages and difficulties. In this investigation, we will dive into the complex scene of innovation, analyzing its positive effects as well as the deterrents it presents.

Benefits:

Upgraded Correspondence:

The appearance of innovation has changed correspondence, separating geological boundaries and empowering immediate connection. From messages and texting to video calls, people and organizations the same advantage from proficient and quick correspondence channels.

Data Availability:

The web fills in as a sweeping store of data, conceding clients admittance to an abundance of information readily available. This democratization of data has engaged people, encouraging schooling, and empowering informed independent direction.

Expanded Efficiency:

Mechanization and advanced devices have fundamentally improved efficiency in different ventures. From assembling cycles to managerial assignments, innovation smoothes out tasks, diminishes physical work, and speeds up yield.

Worldwide Network:

Innovation has worked with worldwide interconnectedness, empowering organizations to extend their span and people to associate with others around the world. Web-based entertainment stages, specifically, play had a significant impact in encouraging worldwide networks and social trade.

Advancement and Inventiveness:

The mechanical scene is a favorable place for development and imagination. Propels in fields like man-made reasoning, augmented experience, and biotechnology prepare for noteworthy disclosures and novel answers for complex issues.

Medical services Progressions:

Innovation has upset medical care, from further developed diagnostics and tele-medicine to the improvement of life-saving clinical gadgets. The joining of information investigation and wearable innovation engages people to proactively screen and deal with their wellbeing.

Monetary Development:

The computerized economy has arisen as a main impetus behind monetary development. Web based business, advanced administrations, and online commercial centers make new roads for organizations, encouraging business venture and occupation creation.

Ecological Protection:

Innovation assumes a urgent part in tending to ecological difficulties. From environmentally friendly power answers for information driven protection endeavors, mechanical advancements add to feasible practices and the safeguarding of regular assets.

Instructive Open doors:

Computerized stages have changed schooling, making learning open to a world-wide crowd. Online courses, digital books, and instructive applications give different and adaptable learning open doors, spanning holes in conventional school systems.

Proficient Asset The executives:

Innovation empowers the productive administration of assets, from savvy lattices improving energy dispersion to information driven approaches in horticulture. These progressions add to maintainability and the capable utilization of assets.

Challenges:

Security Concerns:

The universality of innovation has raised huge worries about security. From information breaks to observation, people and associations face the test of protecting delicate data in an undeniably interconnected world.

Network protection Dangers:

As innovation propels, so do the refinement and recurrence of digital dangers. Malware, phishing assaults, and hacking episodes present dangers to individual information, corporate data, and, surprisingly, basic foundation, requesting powerful network safety measures.

Work Uprooting and Disparity:

Computerization and man-made brainpower can possibly supplant specific positions, prompting worries about joblessness and financial imbalance. Endeavors should be made to reskill the labor force and address inconsistencies emerging from mechanical progressions.

Advanced Enslavement:

The inescapable utilization of computerized gadgets has led to worries about enslavement and its effect on emotional wellness. The steady network and the draw of online entertainment can add to uneasiness, wretchedness, and a feeling of detachment.

E-Squander and Natural Effect:

The fast speed of innovative outdated nature adds to the developing issue of electronic waste. Appropriate removal and reusing of outdated gadgets are fundamental to moderate the natural effect of disposed of hardware.

Counterfeit News and Falsehood:

The advanced time has seen the multiplication of phony news and deception, worked with by the quick scattering of data through web-based stages. This represents a danger to public talk, a vote based system, and social union.

Reliance and Unwavering quality Issues:

Society's rising reliance on innovation makes weaknesses in case of framework disappointments, digital assaults, or cataclysmic events. Guaranteeing the dependability and versatility of innovative framework is pivotal for cultural steadiness.

Moral Problems in Innovation:

The improvement of cutting edge innovations brings up moral issues with respect to their utilization. Issues, for example, artificial intelligence predisposition, independent weapons, and hereditary control require insightful thought of the moral ramifications of innovative progressions.

Social Seclusion:

While innovation interfaces individuals around the world, it can likewise add to social disengagement on a singular level. Unnecessary screen time and virtual communications might supplant up close and personal associations, affecting the nature of connections.

Administrative Difficulties:

The quick speed of mechanical development frequently dominates administrative structures, prompting difficulties in administration and responsibility. Finding some kind of harmony between encouraging advancement and guaranteeing capable use requires versatile and ground breaking administrative measures.

In the embroidered artwork of advantages and difficulties woven by innovation, it is obvious that the effect is sweeping and complex. The advantages, going from further developed correspondence to medical care progressions, have the ability to emphatically change social orders. In any case, the difficulties, from protection worries to work removal, request cautious thought and proactive measures.

As we explore the steadily developing scene of innovation, basic to figure out some kind of harmony expands the positive effect while relieving the unfortunate results. This includes mechanical development as well as insightful guideline, moral contemplations, and a pledge to tending to the cultural ramifications of progressions.

Eventually, the excursion into the fate of innovation is an aggregate undertaking that requires cooperation between people, organizations, states, and the worldwide local area. By saddling the advantages and tending to the difficulties, we can shape a future where innovation fills in as a power for progress, correspondence, and the improvement of humankind.

4.1 Analysis of the advantages of lift irrigation for agricultural productivity

Lift water system is a technique for water supply to horticultural fields that includes the lifting of water from a lower source to a higher rise, commonly utilizing siphons or other mechanical gadgets. This method has acquired unmistakable quality in horticultural practices because of its capability to improve efficiency by guaranteeing a solid and controlled water supply. In this examination, we will investigate the heap benefits of lift water system and its suggestions for agrarian efficiency.

1. **Water Availability and Unwavering quality:**

    One of the essential benefits of lift water system is its capacity to give dependable admittance to water to agrarian exercises. By lifting water from a lower source, ranchers can beat the restrictions forced by the normal geography, guaranteeing a steady and trustworthy water supply for their yields.

2. **Expanded Editing Force:**

    Lift water system empowers ranchers to build the power of editing by considering different trimming seasons in a year. The capacity to control water dissemination and timing upgrades adaptability, empowering the development of harvests with fluctuating water prerequisites consistently.

3. **Moderation of Water Shortage:**

    In locales confronting water shortage, lift water system turns into a urgent device for relieving the effect of lacking precipitation. It furnishes ranchers with the necessary resources to water their fields in any event, during dry periods, lessening the weakness of yields to dry spell conditions.

4. **Further developed Harvest Yield and Quality:**

    The controlled and dependable water supply worked with by lift water system adds to further developed crop yield and quality. Crops get sufficient dampness, which is especially crucial during basic development stages, prompting better plants and better gathers.

5. **Upgraded Land Use:**

    Lift water system takes into consideration the development of terrains that would somehow or another be unsatisfactory for farming because of their rise or distance from normal water sources. This advancement of land use extends the rural impression and adds to expanded food creation.

6. **Productive Water Appropriation:**

    The utilization of lift water system frameworks takes into consideration effective dispersion of water across fields. This is especially gainful in huge scope horticulture, where accuracy in water application is fundamental to stay away from wastage and guarantee that each plant gets the necessary measure of water.

7. **Energy Productivity:**

    Present day lift water system frameworks frequently integrate energy-proficient innovations, like sun based fueled siphons and high level water the board frameworks. This decreases the ecological effect as well as makes lift water system all the more monetarily practical by limiting functional expenses.

8. **Advancement of Practical Agribusiness:**
Lift water system adds to economical horticultural practices by empowering ranchers to take on water-effective water system strategies. By lessening dependence on downpour took care of horticulture, it helps in the protection of water assets and advances harmless to the ecosystem cultivating rehearses.

9. **Social and Financial Turn of events:**
The execution of lift water system activities can have more extensive financial effects. It can prompt expanded business open doors in the agribusiness area, animate nearby economies, and add to generally speaking provincial advancement by improving the vocations of cultivating networks.

10. **Variation to Environmental Change:**
Notwithstanding changing climatic circumstances and eccentric weather conditions, lift water system gives a way to ranchers to adjust to these difficulties. The capacity to control water dispersion becomes urgent in dealing with the effects of environmental change on horticulture.

11. **Decrease of Reliance on Groundwater:**
Lift water system frameworks can diminish the dependence on groundwater for agrarian water system. By taking advantage of surface water sources and effectively using them, ranchers can add to the supportable administration of groundwater assets, forestalling over-extraction and exhaustion.

12. **Disintegration Control:**
The controlled arrival of water in lift water system frameworks can assist in forestalling with ruining disintegration. By keeping a consistent and estimated progression of water, the gamble of soil disintegration is limited, protecting the uprightness of rural land and forestalling the deficiency of fruitful dirt.

13. **Adaptability in Harvest Determination:**
Lift water system gives ranchers the adaptability to pick different harvests in view of market interest, as opposed to being confined by water accessibility. This flexibility adds to monetary broadening and can improve the monetary strength of cultivating networks.

14. **Administrative Help and Speculation:**
Many lift water system projects get backing and venture from legislative bodies. This connotes the acknowledgment of the significance of lift water system in accomplishing agrarian supportability and food security objectives, further reassuring its reception.

15. **Innovative Progressions:**

Progressing headways in innovation keep on improving the productivity and adequacy of lift water system frameworks. Robotized control frameworks, sensor innovations, and information examination add to accuracy agribusiness, advancing water use and expanding generally speaking efficiency.

Taking everything into account, the benefits of lift water system for rural efficiency are complex. From guaranteeing water openness and dependability to moderating water shortage, lift water system assumes a vital part in modernizing and enhancing farming practices.

The capacity to control water dispersion, combined with mechanical developments and supportable practices, positions lift water system as a vital supporter of food security, monetary turn of events, and natural preservation. As horticulture keeps on confronting difficulties presented by a developing worldwide populace and changing climatic circumstances, lift water system stands apart as a significant instrument chasing a stronger and maintainable future for cultivating networks all over the planet.

4.2 Discussion of the challenges and limitations faced by lift irrigation projects

Lift water system projects, while offering various benefits for rural efficiency, are not without their arrangement of difficulties and impediments. In this broad conversation, we will dive into the different intricacies and obstacles looked by lift water system drives, revealing insight into the variables that can hinder their viability.

1. **High Starting Capital Speculation:**

   One of the essential difficulties looked by lift water system projects is the high starting capital speculation expected for their foundation. The expenses related with the development of framework, establishment of siphons, and execution of water dissemination frameworks can be a huge hindrance, especially for limited scope ranchers or asset compelled locales.

2. **Functional and Support Expenses:**

   Past the underlying venture, lift water system projects involve progressing functional and support costs. Normal upkeep of siphons, pipelines, and different parts is fundamental to guarantee the framework's proficiency. The monetary weight of support can strain the financial suitability of lift water system, particularly in regions with restricted monetary assets.

3. **Energy Reliance:**

   Lift water system is in many cases subject to a persistent and solid energy supply, particularly when electric or diesel-controlled siphons are utilized. In districts where power is temperamental or costly, or where fuel accessibility is a worry, the reliance on energy can represent a critical test. This reliance additionally has natural ramifications, as traditional energy sources add to fossil fuel byproducts.

4. **Weakness to Blackouts:**

   Lift water system frameworks are helpless to disturbances in power supply, which can be brought about by variables, for example, network disappointments, gear glitches, or catastrophic events. Blackouts can prompt breaks in water supply, influencing the timing and amount of water system, and possibly compromising harvest wellbeing and yield.

5. **Restricted Water Source Accessibility:**

   The viability of lift water system is dependent upon the accessibility of a solid water source at a lower rise. In districts where such water sources are scant or overexploited, the achievability of lift water system might be compromised. Contest for water assets among different areas, including agribusiness, industry, and homegrown use, further compounds this constraint.

6. **Ecological Effect:**

   The ecological effect of lift water system tasks can be a reason to worry. The modification of normal water streams, especially while redirecting water from streams or other water bodies, can disturb environments and adversely influence oceanic life. Furthermore, the extraction of groundwater for lift water system might add to the exhaustion of springs and fuel water shortage issues.

7. **Land Procurement and Restoration:**

   The execution of lift water system projects frequently includes the procurement of land, which can be a mind boggling and hostile cycle. Uprooting of networks, interruption of jobs, and the requirement for recovery measures should be painstakingly figured out how to guarantee the social and financial prosperity of impacted populaces.

8. **Restricted Inclusion and Scale:**

   Now and again, lift water system activities might be restricted in their inclusion and scale. The framework expected for lift water system, including siphons, pipelines, and conveyance organizations, may just be monetarily reasonable for specific regions or bigger rural activities. This can leave more modest and more remote cultivating networks without admittance to the advantages of lift water system.

9. **Mechanical Intricacy:**

   The intricacy of lift water system advances can present difficulties, particularly in districts where there is an absence of specialized mastery or gifted work. Legitimate establishment, support, and investigating of the hardware require particular information, and the shortfall of such skill can upset the successful activity of lift water system frameworks.

10. **Water Quality Worries:**

    Lift water system frameworks might experience difficulties connected with water quality, particularly while drawing water from streams or different sources with possible defilement. The presence of contaminations, residue, or saltiness in the water can unfavorably influence soil wellbeing and yield efficiency, requiring extra measures like water treatment or soil changes.

11. **Environment Inconstancy and Unusualness:**

    Environment changeability, remembering changes for precipitation examples, dry seasons, or outrageous climate occasions, can present critical difficulties to lift water system projects. Unusual weather patterns can influence the accessibility of water sources and upset the booking of water system, influencing crop development and yield.

12. **Lawful and Administrative Imperatives:**
Lift water system projects should explore a mind boggling trap of lawful and administrative structures. Acquiring fundamental grants, consenting to natural guidelines, and tending to land-use approaches can be tedious and administrative cycles that add layers of intricacy to project execution.

13. **Social and Social Contemplations:**
Lift water system activities might confront opposition or difficulties established in friendly and social elements. Nearby people group might have customary practices connected with water use, and the presentation of new advances or changes in water circulation might be met with wariness or resistance.

14. **Restricted Water Use Proficiency:**
In spite of its benefits, lift water system may not necessarily in every case accomplish ideal water use effectiveness. Wasteful water application strategies, like flooding or lopsided dissemination, can bring about water wastage and diminished viability in advancing yield development. Further developing water use proficiency requires cautious preparation and the executives.

15. **Deficient Observing and Information The board:**
Compelling activity and upkeep of lift water system frameworks depend on exact checking and information the executives. Now and again, insufficient foundation for checking water stream, siphon execution, and soil dampness levels can hinder the convenient identification of issues and the execution of restorative measures.

16. **Absence of Rancher Mindfulness and Preparing:**
The progress of lift water system projects depends on the mindfulness and comprehension of ranchers with respect to the innovation and its advantages. Deficient preparation and mindfulness projects can prompt sub-par use of the framework, restricting its effect on rural efficiency.

17. **Long Execution Timetables:**
The preparation and execution of lift water system projects frequently require expanded courses of events. Defers in project execution, whether because of administrative cycles, financing limitations, or unexpected difficulties, can delay the time before ranchers can profit from the framework.

18. **Monetary Practicality and Profit from Speculation:**

Evaluating the financial feasibility and profit from speculation of lift water system projects is a complicated undertaking. The time expected for the framework to show its advantages, combined with vulnerabilities in crop costs and economic situations, can impact the apparent financial achievability of the undertaking.

In the fantastic embroidery of farming turn of events, lift water system projects stand as a groundbreaking power, bringing the commitment of solid water supply and improved efficiency. Notwithstanding, similarly as with any mechanical intercession, the way to progress is loaded with difficulties and constraints. From monetary obstacles to natural worries, lift water system projects require cautious route through a horde of intricacies.

Tending to these difficulties requests a comprehensive methodology that thinks about the specialized parts of lift water system as well as the social, financial, and ecological aspects. Cooperative endeavors including government bodies, non-administrative associations, nearby networks, and farming partners are fundamental to beating these obstacles and understanding the maximum capacity of lift water system in supporting horticultural occupations and guaranteeing food security. The development of lift water system rehearses should be directed by a promise to supportability, inclusivity, and the prosperity of both the land and the networks it serves.

4.3 Highlighting successful implementations and lessons learned

In the domain of lift water system, various examples of overcoming adversity act as encouraging signs, exhibiting the positive effect of first rate projects on rural efficiency and local area improvement. These examples of overcoming adversity additionally bestow important illustrations that can direct future executions and add to the feasible administration of water assets. Allow us to investigate a few vital instances of fruitful lift water system projects and the examples gained from their accomplishments.

1. **Contextual analysis: Sardar Sarovar Venture, India:**
   The Sardar Sarovar Task, situated in the Indian province of Gujarat, remains as one of the most broad lift water system projects worldwide. The task includes the development of an enormous dam on the Narmada Waterway, making a supply that works with lift water system for farming grounds. By tackling the force of the dam, water is lifted to raised channels, giving water system to immense agrarian regions.

   Illustrations Learned:

   Coordinated Water The executives: The progress of the Sardar Sarovar Venture lies in its far reaching way to deal with water the board. By incorporating dam development, supply creation, and lift water system, the task upgrades water assets for both horticultural and homegrown purposes.

   Local area Association: The venture underscores local area inclusion and cooperation, guaranteeing that nearby ranchers benefit straightforwardly from the water system offices. This model of cooperation encourages a feeling of pride among the cultivating local area and advances the manageable utilization of water assets.

2. **Contextual investigation: Ramthal Marol Lift Water system Undertaking, India:**

The Ramthal Marol Lift Water system Venture, situated in the province of Karnataka, India, is an extraordinary drive pointed toward giving dependable water supply to dry spell inclined regions. The task lifts water from the Krishna Stream to inundate north of 24,000 hectares of rural land, helping various little and minor ranchers.

Examples Learned:

Accuracy Farming Practices: The venture integrates accuracy horticulture works on, streamlining water use through proficient water system procedures. Trickle and sprinkler water system frameworks are utilized, diminishing water wastage and improving harvest yields while guaranteeing water preservation.

Environment Strong Arrangements: Perceiving the weakness of the locale to environment changeability, the Ramthal Marol project integrates environment versatile arrangements. The capacity to adjust to changing atmospheric conditions guarantees the manageability of rural practices notwithstanding environment related difficulties.

3. **Contextual investigation: Israel's Public Water Transporter:**

Israel's Public Water Transporter is a spearheading lift water system project that plays had a critical impact in changing bone-dry scenes into useful farming districts. The task includes the lifting of water from the Ocean of Galilee to inundate fields the nation over, supporting Israel's farming, which appearances water shortage challenges.

Illustrations Learned:

Water Reusing and Reuse: Israel's prosperity lies in its creative way to deal with water reusing and reuse. Treated wastewater is used for rural water system, limiting the interest on freshwater sources and advancing maintainable water the executives rehearses.

Effective Water Conveyance Organizations: The execution of proficient water dispersion organizations, including progressed dribble water system frameworks, guarantees that water arrives at crops unequivocally where and when it is required. This enhancement adds to expanded water use proficiency and horticultural efficiency.

4. **Contextual analysis: Lift Water system in China's Loess Level:**

China's Loess Level, described by testing geography and soil disintegration, has seen effective lift water system projects pointed toward changing fruitless land into useful rural regions. These tasks include lifting water from streams to flood terraced fields, cultivating supportable farming in the district.

Illustrations Learned:

Soil Protection Procedures: The outcome in the Loess Level is credited not exclusively to lift water system yet in addition to the execution of soil preservation methods.

Terracing and form furrowing assist with forestalling soil disintegration, protecting prolific dirt and supporting rural efficiency.

Environment Reclamation: The ventures in the Loess Level exhibit the significance of thinking about more extensive biological system rebuilding. Past water system, drives that address soil wellbeing, vegetation cover, and watershed the executives add to the general manageability of agrarian practices.

5. **Contextual analysis: Lift Water system in the Nile Delta, Egypt:**
The Nile Delta in Egypt has carried out lift water system as a feature of a broad horticultural improvement methodology. The lifting of water from the Nile Stream upholds the water system of fruitful delta lands, contributing fundamentally to Egypt's rural result.

Illustrations Learned:

Verifiable Information Combination: Outcome in the Nile Delta is grounded in the joining of authentic information and customary water system rehearses. By expanding upon extremely old strategies and adjusting them to present day advancements, the task guarantees the coherence of supportable agrarian practices.

Cooperation with Worldwide Accomplices: The venture grandstands the advantages of joint effort with global accomplices. By utilizing aptitude, innovation, and monetary help from outer elements, the lift water system drives in the Nile Delta have had the option to upgrade their adequacy and effect.

6. **Cross-Cutting Illustrations Learned:**

While each contextual investigation presents extraordinary illustrations, there are normal subjects that cut across fruitful lift water system projects:

Comprehensive Preparation: Effective ventures underline all encompassing arranging that thinks about the specialized parts of lift water system as well as more extensive ecological, social, and financial aspects. Thorough arranging guarantees that the venture lines up with the requirements and setting of the district.

Versatile Administration: Adaptability and versatile administration are vital to tending to unanticipated difficulties. Fruitful activities consolidate systems for observing and adjusting to changes in environment, water accessibility, and other unique variables.

Local area Cooperation: Including nearby networks in the preparation, execution, and the board of lift water system projects cultivates a feeling of responsibility and guarantees that the advantages arrive at the planned recipients. Local area commitment adds to the drawn out manageability of the drives.

Mechanical Advancement: Embracing mechanical development is pivotal for the outcome of lift water system projects. The combination of present day water system methods, energy-productive siphons, and accuracy agribusiness rehearses improves the proficiency and supportability of water use.

Limit Building: Giving preparation and building the limit of neighborhood partners, including ranchers and professionals, is crucial. Outfitting people with the information and abilities vital for the activity and upkeep of lift water system frameworks guarantees the life span of the tasks.

In the display of lift water system drives, the examples of overcoming adversity referenced above offer priceless experiences and illustrations that rise above geological limits. The contextual analyses feature the extraordinary capability of lift water system in improving horticultural efficiency, guaranteeing water security, and cultivating supportable turn of events.

The illustrations gained from these fruitful executions highlight the significance of a diverse methodology that incorporates specialized development, ecological stewardship, local area commitment, and versatile administration. As we plan ahead, these examples act as guideposts for policymakers, project implementers, and networks trying to tackle the advantages of lift water system while exploring the difficulties intrinsic in overseeing water assets for agribusiness.

Obviously the outcome of lift water system projects not entirely settled by the specialized complexities of siphoning water to raised landscapes however by the capacity to blend with the common habitat, elevate nearby networks, and adjust to the advancing elements of environment and horticulture. In winding around together these components, lift water system can possibly arise as a mechanical arrangement as well as an impetus for reasonable and strong farming scenes all over the planet.

The execution of different tasks across various areas and spaces has yielded significant illustrations that stretch out past the particular setting of every drive. These illustrations learned give experiences into viable methodologies, traps to keep away from, and best practices that can illuminate future undertakings. In this investigation, we will dig into assorted executions and the general examples they offer.

1. **Efficient power Energy Change:**

   In the domain of efficient power energy change, the sending of environmentally friendly power projects, for example, sun based and wind ranches, has been a point of convergence. These drives intend to move energy creation away from petroleum products, adding to manageability and moderating the effects of environmental change.

   Illustrations Learned:

   Strategy Systems: Effective environmentally friendly power energy changes frequently depend on strong arrangement structures that give motivating forces to sustainable power reception.

   States assume a urgent part in establishing a climate helpful for interest in clean energy by offering endowments, charge motivations, and administrative help.

Local area Commitment: Connecting with neighborhood networks is crucial for the outcome of sustainable power projects. Tending to local area concerns, guaranteeing fair remuneration for land use, and including occupants in the dynamic cycle cultivate a feeling of responsibility and decrease resistance to projects.

Mechanical Advancement: Consistent mechanical development is critical to the progress of environmentally friendly power energy projects. Propels in sun powered charger productivity, energy capacity arrangements, and framework the board advancements add to the financial feasibility and versatility of sustainable power drives.

2. **Savvy City Drives:**

The idea of savvy urban areas includes utilizing innovation to upgrade metropolitan living, further develop effectiveness, and advance maintainability. Shrewd city drives incorporate different components, including advanced framework, insightful transportation frameworks, and information driven administration.

Examples Learned:

Information Protection and Security: As shrewd city drives depend vigorously on information, guaranteeing powerful information protection and safety efforts is vital. Implementers should focus on the insurance of resident information, address network protection concerns, and lay out straightforward conventions for information utilization.

Cooperative Biological systems: Fruitful shrewd urban areas frequently rise out of cooperative environments including government organizations, confidential ventures, the scholarly world, and residents. Building organizations works with asset sharing, advancement, and the improvement of incorporated arrangements that address complex metropolitan difficulties.

Comprehensive Plan: The plan of savvy city advances ought to focus on inclusivity, taking into account the different necessities of the populace. Openness highlights, moderateness contemplations, and fair admittance to advanced administrations guarantee that the advantages of brilliant city drives are shared by all occupants.

3. **Medical services Innovation Reception:**

The reconciliation of innovation in medical care has seen the execution of electronic wellbeing records, telemedicine, and creative clinical gadgets. These drives expect to work on understanding consideration, upgrade diagnostics, and smooth out medical care conveyance.

Illustrations Learned:

Interoperability Norms: Guaranteeing interoperability between various medical care frameworks and innovations is basic for powerful data trade. Normalized conventions and information designs empower consistent correspondence between assorted medical services applications, lessening discontinuity and working on quiet consideration.

Client Focused Plan: Medical services innovations should be planned with an emphasis on the end-clients, including medical care experts and patients. Client focused plan standards upgrade ease of use, limit blunders, and further develop in general client fulfillment, at last prompting better medical services results.

Administrative Consistence: Consistence with administrative guidelines is non-debatable in the medical care area. Complying with guidelines, for example, the Medical coverage Movability and Responsibility Act (HIPAA) guarantees the protection and security of patient data, imparting trust in innovation reception.

4. **E-Learning Stages:**

The execution of e-learning stages has become progressively predominant, particularly right after worldwide changes in training conveyance. Internet learning drives expect to give available and adaptable schooling open doors, rising above geological limits.

Examples Learned:

Advanced Incorporation: Guaranteeing computerized consideration is vital for the progress of e-learning drives. Crossing over the computerized partition by giving admittance to innovation, web network, and important gadgets is fundamental to keep minimized populaces from being abandoned.

Commitment Systems: Compelling e-learning goes past satisfied conveyance; it includes connecting with understudies in a virtual climate. Implementers ought to utilize intelligent substance, gatherings, and cooperative instruments to cultivate understudy commitment, inspiration, and a feeling of local area.

Adaptability and Adaptability: Versatility and adaptability are principal in e-learning stages, particularly taking into account possible expansions in client numbers. Cloud-based arrangements, versatile learning innovations, and measured content plan add to adaptability and flexibility.

5. **Monetary Consideration Drives:**

Monetary incorporation drives mean to carry banking and monetary administrations to underserved populaces. Executions incorporate versatile banking, microfinance, and inventive fintech arrangements intended to address hindrances to monetary access.

Examples Learned:

Client Instruction: Teaching clients about monetary items and administrations is a basic part of monetary consideration. Numerous people in underserved networks might be new to banking ideas, and thorough client training programs improve monetary proficiency.

Administrative Help: A steady administrative climate is fundamental for the outcome of monetary consideration drives. States and administrative bodies assume a key part in making structures that energize development, safeguard buyers, and cultivate the development of comprehensive monetary administrations.

Associations for Last-Mile Access: Laying out associations with neighborhood substances, including local area associations and portable organization administrators, works with last-mile access. Utilizing existing foundations and working together with nearby partners improves the compass and effect of monetary consideration projects.

6. **Fiasco Reaction and Readiness:**

Executions in misfortune reaction and readiness include the utilization of innovation and information to upgrade early advance notice frameworks, coordination endeavors, and strength working in networks helpless against catastrophic events.

Illustrations Learned:

Local area Based Approaches: Effective debacle reaction drives frequently include local area based approaches that engage neighborhood inhabitants. Local area commitment, preparing in calamity readiness, and the foundation of local area networks add to quick and compelling reactions.

Information Examination for Early Advance notice: The utilization of information investigation, including satellite symbolism and weather conditions conjectures, empowers early admonition frameworks. Implementers can use prescient examination to expect catastrophes, giving networks additional opportunity to plan and clear if important.

Cross-Sectoral Coordinated effort: Fiasco reaction is intrinsically cross-sectoral, requiring cooperation between government offices, non-legislative associations, confidential elements, and neighborhood networks. Laying out coordination instruments and data sharing conventions upgrades the viability of calamity reaction endeavors.

7. **Space Investigation and Satellite Innovation:**

Space investigation and satellite innovation executions have extended how we might interpret the universe and added to headways in correspondence, weather conditions guaging, and Earth perception.

Illustrations Learned:

Worldwide Coordinated effort: Space investigation frequently includes global cooperation, uniting the aptitude and assets of various nations. Cooperative space missions, information sharing arrangements, and joint examination endeavors add to the progress of room investigation drives.

Innovative Side projects: Advancements produced for space investigation frequently have applications past their unique reason. Advancements like satellite correspondence, GPS innovation, and materials produced for space missions have tracked down far reaching use in different enterprises on The planet.

Long haul Arranging and Speculation: Fruitful space investigation drives require long haul arranging and supported venture. The improvement of rocket, satellite frameworks, and logical instruments requests huge monetary responsibilities and a guarantee to persistent investigation and revelation.

8. **Examples Gained from Cross-Cutting Subjects:**

Across these different executions, a few cross-cutting subjects arise, giving overall examples that are material to many undertakings:

Versatile Initiative: Fruitful executions are many times driven by versatile pioneers who can explore vulnerability, embrace change, and encourage advancement. Versatile initiative is fundamental for controlling ventures through unexpected difficulties and guaranteeing long haul supportability.

Moral Contemplations: Moral contemplations ought to be implanted in project plan and execution. This incorporates focusing on client protection, guaranteeing fair access, and tending to expected social, monetary, and ecological effects.

Observing and Assessment: Thorough checking and assessment processes are basic for surveying project influence, recognizing regions for development, and guaranteeing responsibility. Implementers ought to lay out obvious signs, gather pertinent information, and direct normal assessments.

Chapter 5

Environmental Impact

The ecological effect of human exercises has turned into a squeezing worry in ongoing many years. As the worldwide populace proceeds to develop and industrialization grows, the regular world appearances extraordinary difficulties. From deforestation to air and water contamination, the results of human activities on the climate are sweeping and complex.

One of the most apparent and quick effects of human action on the climate is deforestation. The getting free from enormous breadths of backwoods for agribusiness, logging, and metropolitan improvement has prompted the deficiency of biodiversity, interruption of environments, and the arrival of huge measures of carbon dioxide into the air. Timberlands assume a urgent part in sequestering carbon and keeping up with the equilibrium of ozone depleting substances. At the point when they are chopped down or debased, the put away carbon is delivered, adding to environmental change.

Talking about environmental change, it is maybe the most basic ecological issue within recent memory. The consuming of petroleum products, like coal, oil, and flammable gas, discharges ozone harming substances, including carbon dioxide and methane, into the climate. These gases trap heat, prompting a steady climb in worldwide temperatures. The outcomes of environmental change are huge and incorporate more successive and extreme climate occasions, rising ocean levels, and interruptions to biological systems and farming.

Notwithstanding environmental change, contamination represents a huge danger to the climate. Air contamination, essentially from the consuming of petroleum derivatives and modern cycles, adds to respiratory issues, brown haze development, and corrosive downpour. Water contamination, brought about by modern releases, farming spillover, and ill-advised garbage removal, hurts oceanic biological systems, imperils marine life, and compromises the accessibility of clean drinking water.

The utilization of pesticides and composts in agribusiness affects the climate. Overflow from fields can convey these synthetic substances into streams and streams, prompting the pollution of water sources. This damages amphibian life as well as have flowing consequences for human wellbeing as these impurities advance up the established pecking order.

The exhaustion of normal assets is one more outcome of human exercises that has significant natural ramifications. From overfishing in the seas to the extraction of minerals and non-renewable energy sources, people are consuming assets at an impractical rate. This compromises the equilibrium of environments as well as risks the drawn out accessibility of fundamental assets for people in the future.

Urbanization and the development of foundation likewise add to natural corruption. As urban communities grow, normal territories are many times supplanted by cement and black-top. This outcomes in the fracture of environments, making it hard for untamed life to track down reasonable living spaces and move. Metropolitan regions likewise create heat islands, where temperatures are higher than in encompassing country regions, further adding to changes in neighborhood environments.

The creation and removal of plastic have arisen as a worldwide ecological emergency. Single-use plastics, like packs and containers, continue in the climate for quite a long time, contaminating seas and hurting marine life. The size of plastic contamination has arrived at such extents that microplastics, small particles coming about because of the breakdown of bigger plastic things, are currently found in the air we inhale, the water we drink, and the food we eat.

The natural effect of human exercises reaches out past the actual domain to incorporate the deficiency of biodiversity.

The fast downfall of plant and creature species is an immediate consequence of environment obliteration, contamination, environmental change, and overexploitation. Biodiversity misfortune not just reduces the excellence of the normal world yet additionally sabotages the strength of environments, making them more defenseless against aggravations and less fit for offering fundamental types of assistance, like fertilization and water purging.

The outcomes of ecological corruption are not uniformly dispersed, with under-estimated networks frequently enduring the worst part of the effects. Ecological equity issues feature the lopsided openness of specific populaces to contamination, risky waste, and the impacts of environmental change. Low-pay networks and networks of variety are much of the time arranged close to modern offices and experience higher paces of medical issues accordingly.

Tending to the natural effect of human exercises requires a far reaching and composed exertion at the nearby, public, and worldwide levels. Economical practices in horticulture, ranger service, and fisheries are fundamental to guarantee the dependable utilization of regular assets. Progressing to environmentally friendly power sources, for example, sun oriented and wind power, is significant for moderating environmental change and decreasing dependence on petroleum derivatives.

The reception of roundabout economy standards, which focus on reusing and limiting waste, can assist with decreasing the ecological effect of creation and utilization. States and organizations assume a significant part in carrying out and boosting maintainable practices, from controlling emanations to advancing eco-accommodating innovations.

Worldwide collaboration is additionally imperative in tending to worldwide natural difficulties. Environmental change, contamination, and biodiversity misfortune are interconnected issues that rise above public boundaries. Cooperative endeavors, for example, the Paris Understanding, look to unite nations to set targets and responsibilities for moderating environmental change and advancing practical turn of events.

Instructing the general population about the ecological effect of their decisions and activities is key to encouraging a feeling of obligation and advancing economical way of behaving. From buyer decisions to way of life propensities, people can add to positive ecological results through informed choices. Public mindfulness crusades, ecological training in schools, and the advancement of reasonable ways of life are fundamental parts of this work.

The combination of nature-based arrangements into metropolitan preparation and configuration can assist with alleviating the effect of urbanization on the climate. Green spaces, reasonable design, and the rebuilding of regular environments inside urban areas add to biodiversity preservation and work on the personal satisfaction for occupants. Shrewd and supportable metropolitan advancement is significant for making versatile and harmless to the ecosystem urban areas.

Mechanical development likewise assumes a key part in tending to natural difficulties. Progresses in clean energy advancements, like electric vehicles, energy capacity, and environmentally friendly power frameworks, offer options to conventional, contaminating sources. The advancement of supportable and eco-accommodating materials, as well as forward leaps in squander the executives and reusing, add to diminishing the natural impression of human exercises.

While endeavors to address natural effect have gained ground in certain areas, huge difficulties remain. The criticalness of the circumstance requires a reconsideration of needs and a change in outlook in how social orders approach improvement. Offsetting monetary development with natural supportability is an intricate undertaking that requires a guarantee to long haul thinking and a readiness to settle on hard choices.

All in all, the ecological effect of human exercises is a complicated and interconnected issue that requires prompt and deliberate activity. From environmental change to contamination and biodiversity misfortune, the outcomes of our activities are clear on a worldwide scale. Accomplishing a manageable and agreeable relationship with the normal world requires a complex methodology, including people, networks, legislatures, and organizations.

Our decisions today will shape the ecological inheritance we leave for people in the future. By embracing feasible works on, upholding for natural security, and cultivating a feeling of obligation, we can make progress toward a stronger and biologically adjusted planet. The difficulties are overwhelming, however the potential for positive change is inside our grip. An aggregate exertion requires the responsibility of all of us to get a supportable and flourishing future for the planet we call home.

5.1 Examination of the environmental implications of lift irrigation

Lift water system, a strategy for water supply in horticulture, has been generally utilized across the globe to address the water needs of yields in regions where normal water accessibility is deficient. While lift water system frameworks assume a critical part in upgrading horticultural efficiency and supporting food security, it is fundamental to extensively look at their natural ramifications. This assessment envelops different features, including water utilization effectiveness, influences on environments, energy utilization, and possible ramifications for nearby networks.

Water shortage is a squeezing worldwide issue, and lift water system frameworks are frequently executed to defeat limits forced by unpredictable water conveyance or lacking precipitation. These frameworks commonly include siphoning water from a water source, like a stream or supply, to higher rises, permitting it to move through channels to inundate fields. While lift water system helps satisfy farming water needs, it is fundamental to evaluate the proficiency of water use in these frameworks.

One of the key ecological contemplations is how much water expected for lift water system and the potential for water wastage. Wasteful water use can prompt over-extraction from water sources, adversely influencing amphibian environments and exhausting groundwater holds. The assessment of lift water system's ecological ramifications ought to include an evaluation of the water extraction rates, the adequacy of water conveyance to fields, and the actions set up to limit water misfortunes during the interaction.

The modification of normal water stream designs is one more perspective that warrants consideration in the assessment of lift water system. The redirection of water from streams or repositories to inundate fields can upset biological systems downstream. This modification might influence sea-going environments, amphibian biodiversity, and the general wellbeing of riverine biological systems. Also, changes in water accessibility downstream can affect networks depending on these water hotspots for drinking water, fishing, and other business exercises.

The energy necessities of lift water system frameworks comprise a critical ecological thought. Most lift water system frameworks depend on siphons fueled by power or diesel motors to lift water to higher heights. The energy sources and proficiency of these siphons assume a pivotal part in deciding the ecological effect of lift water system. The utilization of petroleum products adds to air contamination and ozone harming substance discharges, worsening environmental change. Subsequently, an intensive assessment ought to incorporate an evaluation of the energy sources, energy proficiency, and carbon impression related with lift water system tasks.

Land use changes coming about because of the execution of lift water system frameworks can have assorted ecological outcomes. Getting land for the establishment free from siphoning stations, channels, and water system foundation might prompt natural surroundings misfortune and discontinuity. The disturbance of regular scenes can influence nearby verdure, possibly prompting a decrease in biodiversity. An assessment of lift water system's ecological ramifications ought to consider the degree of land use changes, their effect on environments, and methodologies to moderate adverse consequences on biodiversity.

Soil disintegration is one more natural concern related with lift water system. The adjusted progression of water and expanded dampness levels in flooded fields can add to soil disintegration. Sedimentation in water bodies downstream can debase water quality and mischief oceanic environments. The assessment of lift water system's natural ramifications ought to address soil disintegration chances, disintegration control measures executed, and the likely downstream effects on water bodies.

The financial parts of lift water system ought to be viewed as pair with its ecological ramifications. Neighborhood people group frequently depend on water assets for different necessities, and changes in water accessibility because of lift water system can have significant outcomes.

An assessment ought to incorporate an evaluation of the social and financial effects on networks, considering issues like changes in water accessibility for homegrown use, expected clashes over water assets, and the general prosperity of neighborhood populaces.

Groundwater consumption is a huge concern related with lift water system, particularly in regions where springs are the essential wellspring of water system water. Over the top siphoning from springs can prompt the bringing down of groundwater levels, influencing the accessibility of water for both horticultural and non-rural purposes. An assessment of lift water system's ecological ramifications ought to dive into the groundwater elements, re-energize components, and the potential for long haul influences on spring supportability.

Carrying out water the board rehearses that advance maintainability is urgent in relieving the natural effect of lift water system. This incorporates the reception of water-productive water system advancements, for example, dribble or sprinkler frameworks, which limit water wastage and improve the accuracy of water application. Furthermore, consolidating current checking and control frameworks can advance water use, decreasing the in general ecological impression of lift water system.

Natural rebuilding endeavors can assist with balancing the ecological effects of lift water system by restoring regions impacted via land use changes. This might include reforestation, wetland rebuilding, and living space improvement to help biodiversity and environment versatility. Practical land use arranging that considers the natural conveying limit of the scene is fundamental in limiting unfavorable ecological outcomes.

Local area commitment and participatory methodologies are essential to the practical execution of lift water system frameworks. Including nearby networks in dynamic cycles, taking into account customary information, and tending to the financial necessities of the populace can add to the general achievement and manageability of lift water system projects. This cooperative methodology encourages a feeling of pride among networks, advancing capable water use and natural stewardship.

Strategy structures and administrative measures assume a critical part in molding the ecological results of lift water system projects. States and administrative bodies ought to lay out rules that focus on ecological manageability, proficient water use, and the insurance of biological systems. Observing and implementation systems are fundamental to guarantee consistence with ecological principles and address any potential adverse consequences emerging from lift water system exercises.

Incorporated water asset the board (IWRM) gives a comprehensive way to deal with address the ecological ramifications of lift water system. IWRM stresses the planned turn of events and the board of water, land, and related assets to boost financial and social government assistance without compromising the supportability of biological systems.

By taking on IWRM standards, chiefs can figure out some kind of harmony between horticultural water needs and ecological protection, advancing an amicable concurrence between human exercises and the regular habitat.

All in all, the assessment of the natural ramifications of lift water system requires a far reaching and multi-faceted methodology. Surveying water use productivity, influences on biological systems, energy utilization, land use changes, soil disintegration gambles, financial outcomes, and groundwater elements is fundamental in understanding the full range of natural difficulties related with lift water system. Carrying out economical works on, consolidating mechanical developments, connecting with nearby networks, and laying out strong approach structures are key parts of moderating the natural effect and guaranteeing the drawn out reasonability of lift water system frameworks. By tending to these viewpoints aggregately, it is feasible to outfit the advantages of lift water system while protecting the environmental uprightness of scenes and water assets.

5.2 Consideration of sustainable practices and mitigating measures

The thought of reasonable practices and alleviating measures is principal in tending to the bunch natural difficulties we face today. As human exercises keep on applying significant effects in the world, from environmental change to biodiversity misfortune, there is a critical need to progress towards rehearses that advance natural flexibility and limit hurt. This assessment will dig into different parts of maintainable practices and alleviating measures across various areas, underscoring the interconnectedness of ecological, social, and monetary prosperity.

In agribusiness, the reception of economical cultivating rehearses is significant for guaranteeing food security while limiting ecological corruption. Agroecological approaches, like natural cultivating, stress the combination of biological standards into horticultural frameworks. This includes rehearses like harvest revolution, cover trimming, and decreased dependence on engineered pesticides and manures. By focusing on soil wellbeing, biodiversity, and water protection, economical horticulture mitigates ecological effects as well as improves the flexibility of cultivating frameworks to environmental change.

Accuracy horticulture, empowered by current advances, for example, GPS-directed work vehicles and robots, adds to manageable cultivating by improving asset use. Accuracy horticulture permits ranchers to tailor data sources like water, manures, and pesticides to explicit region of a field, decreasing waste and limiting natural effect. The coordination of information driven dynamic in horticulture upgrades proficiency and efficiency while limiting the natural impression of cultivating activities.

Water the board is a basic part of feasible practices, especially in districts confronting water shortage. The execution of water-saving advances, for example, trickle water system and water reaping, assists preserve with watering assets and improve rural manageability.

Moreover, the reclamation and security of normal water environments, for example, wetlands and watersheds, assume a crucial part in keeping up with water quality and controlling water accessibility for both farming and non-horticultural purposes.

In the energy area, the progress to environmentally friendly power sources is a basic moderating measure to address environmental change and decrease reliance on petroleum products. Sunlight based, wind, hydroelectric, and geothermal power age offer cleaner options with lower ozone depleting substance discharges. The turn of events and organization of energy stockpiling innovations, like high level batteries, add to the security and dependability of environmentally friendly power frameworks, empowering a more reasonable and versatile energy foundation.

Energy productivity gauges likewise assume a crucial part in relieving ecological effect. Enhancements in building plan, machine proficiency, and modern cycles add to decreased energy utilization. Government arrangements and impetuses that advance energy proficiency, alongside open mindfulness crusades, urge people and organizations to take on supportable practices in their energy use.

The transportation area presents the two difficulties and amazing open doors for supportability. The charge of vehicles, combined with headways in battery innovation, adds to diminishing outflows from the transportation area. Putting resources into public transportation framework, advancing strolling and cycling, and creating savvy metropolitan arranging add to supportable versatility arrangements that diminish gridlock and air contamination.

The idea of roundabout economy standards is getting some momentum as a key moderating measure across different enterprises. Rather than the conventional straight model of creation and utilization, where items are produced, utilized, and disposed of, a roundabout economy looks to limit squander and expand asset productivity.

Reusing, reusing, and revamping items and materials add to a more manageable and harmless to the ecosystem way to deal with creation and utilization.

Squander the executives rehearses are basic to alleviating natural effects, particularly with regards to the developing worldwide waste issue. Legitimate garbage removal, reusing, and the decrease of single-use plastics add to the safeguarding of biological systems, decrease of contamination, and preservation of assets. Squander to-energy advances give imaginative answers for tackle energy from squander materials, further adding to economical waste administration.

The preservation and reclamation of regular biological systems are fundamental relieving measures against biodiversity misfortune and territory obliteration. Safe-guarded regions, untamed life passageways, and reforestation endeavors add to the conservation of biodiversity and the improvement of environment administrations.

Native and neighborhood local area association in protection drives guarantees the supportable administration of normal assets while regarding conventional information and practices.

In metropolitan preparation and advancement, the idea of practical urban communities accentuates establishing metropolitan conditions that focus on ecological maintainability, social inclusivity, and monetary suitability. Green structure rehearses, energy-effective foundation, and the coordination of green spaces add to the production of urban communities that limit their natural effect and upgrade the personal satisfaction for occupants. Maintainable metropolitan transportation, squander the executives, and water protection are key parts of building tough and eco-accommodating metropolitan spaces.

The thought of supportable practices reaches out to the corporate area, where organizations assume a huge part in forming ecological results. Corporate social obligation (CSR) drives that emphasis on natural supportability, moral obtaining, and local area commitment add to a positive effect on both the climate and society. Green inventory network rehearses, maintainable creation processes, and eco-accommodating bundling are becoming fundamental parts of capable strategic policies.

The idea of eco-marking furnishes customers with data about the natural effect of items, empowering them to settle on educated and maintainable decisions. Accreditations like Fair Exchange and Timberland Stewardship Gathering (FSC) affirmation guarantee that items are obtained and delivered in an ecologically and socially capable way. Shopper mindfulness and interest for supportable items drive organizations towards taking on harmless to the ecosystem rehearses all through their inventory chains.

Schooling and mindfulness crusades are fundamental relieving measures to cultivate a more extensive comprehension of natural issues and advance supportable way of behaving. Natural training in schools, public effort projects, and media crusades add to building a general public that qualities and effectively partakes in manageable practices. By bringing issues to light about the interconnectedness of natural, social, and financial frameworks, people can pursue informed decisions that add to positive ecological results.

Government arrangements and global collaboration are imperative in driving foundational change and guaranteeing the boundless reception of supportable practices. Guidelines that set ecological norms, boost manageable practices, and punish natural corruption give a system to organizations and people to dependably work. Peaceful accords, for example, the Paris Settlement on environmental change, show the worldwide obligation to tending to ecological difficulties through cooperative endeavors.

Monetary components, for example, green supporting and reasonable speculation rehearses, add to coordinating capital towards harmless to the ecosystem projects.

Financial backers progressively think about ecological, social, and administration (ESG) factors in their navigation, boosting organizations to take on feasible practices and straightforward detailing. The coordination of maintainability rules into monetary direction adds to the arrangement of financial exercises with natural objectives.

Mechanical advancement keeps on assuming an essential part in creating economical arrangements and relieving ecological effect. Propels in clean energy advances, maintainable horticulture practices, and waste administration advancements add to an additional harmless to the ecosystem and strong future. Innovative work endeavors zeroed in on supportable advances can possibly change businesses and advance the reception of eco-accommodating practices.

All in all, the thought of reasonable practices and moderating measures is basic in exploring the mind boggling difficulties presented by natural debasement. From farming to energy, transportation to squander the executives, and corporate practices to individual decisions, supportability should be at the front of direction. The mix of ecological contemplations into all parts of human movement isn't just an ethical objective yet in addition a down to earth way to deal with guaranteeing the drawn out prosperity of the planet and its occupants. By embracing reasonable practices and carrying out moderating measures, we can manufacture a way towards a stronger, evenhanded, and ecologically manageable future.

5.3 Exploration of the balance between agricultural needs and ecological concerns

The investigation of the sensitive harmony between rural necessities and natural worries is a basic undertaking as the world wrestles with taking care of a developing populace while endeavoring to support the wellbeing and flexibility of environments. Farming, as an essential human action, assumes a focal part in molding scenes, impacting biodiversity, and affecting regular assets. Accomplishing an amicable harmony between satisfying the food needs of a blossoming worldwide populace and defending the climate requires a nuanced assessment of different variables and the execution of maintainable practices.

Agribusiness has gone through huge changes throughout the long term, driven by innovative headways, populace development, and changing dietary inclinations. The

Green Upheaval, which started during the twentieth 100 years, denoted a time of expanded farming efficiency using high-yielding harvest assortments, manufactured manures, and pesticides. While the Green Upheaval prevailed with regards to supporting food creation and lightening hunger in many areas of the planet, it likewise achieved potentially negative results for the climate.

The broad utilization of engineered inputs in agribusiness, like substance composts and pesticides, has raised environmental worries.

Overflow from farming fields containing these synthetics can taint water sources, prompting water contamination and adversely affecting sea-going environments. Pesticides, intended to control bothers that undermine crops, can have potentially negative side-effects by hurting non-target species, including useful bugs, birds, and sea-going life forms. The investigation of the harmony between farming necessities and biological worries requires a reexamination of these regular practices.

Lately, there has been a developing acknowledgment of the requirement for additional economical and harmless to the ecosystem rural practices. Economical horticulture looks to coordinate environmental standards into cultivating frameworks, stressing soil wellbeing, biodiversity preservation, and asset effectiveness. Agroecology, a methodology established in environmental standards, advances the plan and the executives of horticultural frameworks that impersonate regular biological systems. By encouraging biodiversity, diminishing dependence on outer data sources, and upgrading soil fruitfulness, agroecology intends to address both farming and natural targets.

One vital part of the investigation of this sensitive equilibrium is soil wellbeing. Soil is a fundamental part of farming environments, giving the establishment to establish development and supplement cycling. Regular farming practices, like monoculture and extreme utilization of synthetic data sources, can debase soil quality over the long run. Soil disintegration, loss of natural matter, and supplement consumption are normal results of unreasonable soil the executives. Maintainable rural works on, including cover trimming, crop revolution, and natural cultivating, center around protecting and upgrading soil wellbeing, advancing long haul farming supportability.

Biodiversity preservation is one more basic thought in the investigation of the harmony among farming and biology. Farming scenes that help a different cluster of plant and animal species are stronger to nuisances, sicknesses, and natural vacillations. Conventional cultivating frameworks, frequently described by polyculture and blended editing, add to biodiversity preservation by making heterogeneous scenes that help different species. Interestingly, huge scope monoculture can prompt the improvement of scenes, making them more defenseless to irritations and illnesses and compromising in general biological strength.

The job of pollinators, for example, honey bees and butterflies, in supporting agrarian efficiency features the interconnectedness of farming and environment. Pollinators assume a significant part in the propagation of many blooming plants, including various harvest species. In any case, the utilization of specific pesticides, territory misfortune, and natural changes compromise pollinator populaces. Supportable agrarian practices expect to secure and advance pollinator wellbeing by limiting pesticide use, giving territory passages, and embracing pollinator-accommodating yield the board methodologies.

Water the board is a focal part of the investigation of the harmony among farming and nature. Farming is a significant customer of water assets, and wasteful water system practices can prompt water shortage, especially in parched and semi-bone-dry locales. The over-extraction of groundwater for water system adds to the consumption of springs and postures long haul difficulties for water supportability. Feasible water the board in agribusiness includes the reception of proficient water system advances, water gathering, and the reclamation of regular water environments.

The protection of regular natural surroundings inside and around rural scenes is fundamental for keeping up with biological equilibrium. Fracture and transformation of normal natural surroundings for farming add to environment misfortune, undermining the endurance of many plant and animal species. Agroforestry, the coordination of trees and bushes into horticultural frameworks, gives a way to ration biodiversity, further develop soil wellbeing, and improve biological system administrations. Riparian cradles and untamed life halls in agrarian scenes add to the safeguarding of normal territories and backing biodiversity.

Hereditary variety in crops is a basic figure building strength to changing ecological circumstances and arising vermin and sicknesses. The far reaching reception of a set number of high-yielding harvest assortments, an outcome of present day modern horticulture, has prompted a decrease in crop hereditary variety. This absence of variety makes farming frameworks more defenseless against shocks and restricts the potential for variation. Investigating the harmony between agrarian necessities and natural worries expects endeavors to advance and moderate yield hereditary variety through drives, for example, seed banks, local area seed trades, and the development of conventional and privately adjusted assortments.

The investigation of this sensitive equilibrium likewise includes a thought of the social and financial components of farming. The livelihoods of millions of individuals rely upon farming, especially in rustic areas of non-industrial nations. Economical agrarian practices shouldn't just address biological worries yet additionally add to the prosperity of cultivating networks. This incorporates elevating impartial admittance to land, assets, and markets, as well as encouraging versatility notwithstanding environmental change and other outer difficulties.

Incorporated bug the executives (IPM) addresses a feasible way to deal with bug control that looks to limit the utilization of substance pesticides while successfully overseeing bug populaces. IPM consolidates organic control strategies, like the utilization of regular hunters and parasites, social practices, and the prudent utilization of substance intercessions when fundamental. By taking into account the environmental elements of bug populaces and their regular foes, IPM intends to keep an equilibrium that upholds both rural efficiency and natural wellbeing.

The idea of regenerative farming makes manageability a stride further by zeroing in on rehearses that effectively add to the recovery of environments. Regenerative farming expects to reconstruct soil wellbeing, improve biodiversity, and sequester carbon in soils. Practices, for example, agroforestry, cover trimming, and rotational brushing are basic to regenerative farming, underscoring an all encompassing methodology that tends to the reliance of horticulture and biology.

Biological system benefits, the advantages that environments give to human social orders, are a significant thought in the investigation of the harmony among horticulture and biology. These administrations incorporate fertilization of yields, water decontamination, soil fruitfulness, and environment guideline. Feasible farming practices that focus on biological system wellbeing add to the proceeded with arrangement of these administrations. Protection horticulture, which limits soil aggravation and advances cover editing, improves soil structure, lessens disintegration, and supports environment administrations basic for farming.

The investigation of the harmony between rural requirements and natural worries should likewise think about the worldwide components of food creation and exchange. The development of agrarian boondocks, frequently to the detriment of normal biological systems like timberlands, has huge ramifications for biodiversity and carbon sequestration. Global exchange agrarian items can prompt the uprooting of natural effects, with the ecological results of creation happening in one locale while the merchandise are consumed in another. Economical horticultural practices ought to represent these worldwide aspects and take a stab at a more evenhanded conveyance of ecological obligations.

The reception of agroecological standards lines up with the investigation of the harmony between farming necessities and biological worries. Agroecology perceives the intricacy of farming frameworks and means to work with regular cycles instead of against them. By underscoring variety, natural strength, and the joining of nearby information, agroecology gives a system to accomplishing supportable food creation while limiting negative ecological effects.

nuanced approach. The sensitive exchange between fulfilling the developing needs of worldwide food creation and protecting the wellbeing and versatility of environments presents complex difficulties that require cautious thought and manageable practices. Farming, as one of the most seasoned and most fundamental human exercises, is unpredictably connected to the common habitat, molding scenes, affecting biodiversity, and influencing the fragile equilibrium of environments. In this investigation, we dive into the multi-layered components of the connection between rural requirements and natural worries, meaning to disentangle the intricacies and distinguish pathways towards a more maintainable conjunction.

The authentic direction of agribusiness has seen groundbreaking shifts, especially with the appearance of the Green Transformation during the twentieth hundred years.

This period denoted a critical expansion in rural efficiency through the far reaching reception of high-yielding harvest assortments, engineered composts, and pesticides. While these developments assumed a urgent part in tending to prompt food security concerns and lifting millions out of neediness, they likewise presented potentially negative results for the climate. The investigation of the harmony between horticultural necessities and natural worries requires a review examination of these past mediations and a basic reassessment of their drawn out supportability.

Manufactured inputs, like substance manures and pesticides, have become meaningful of traditional horticulture. In any case, their broad use raises biological worries that request examination. Overflow from horticultural fields, weighed down with these synthetics, represents a danger to water sources, adding to water contamination and unfavorably influencing oceanic biological systems. The unseen side-effects of pesticide use, including mischief to non-target species and the advancement of safe vermin populaces, highlight the natural intricacy that rural practices should wrestle with. As we investigate this sensitive equilibrium, it becomes clear that a shift towards additional maintainable and environmentally careful practices is basic.

The worldview of reasonable horticulture arises as a core value in exploring the nexus between farming requirements and natural worries. Reasonable farming tries to blend the objectives of food creation with natural wellbeing, recognizing the multifaceted snare of connections inside biological systems. Key to this approach is the idea of agroecology, which coordinates environmental standards into cultivating frameworks. Agroecological rehearses focus on soil wellbeing, biodiversity preservation, and asset productivity, meaning to manufacture a more cooperative connection among farming and nature.

Soil wellbeing remains as an essential point of support in the investigation of this fragile equilibrium. Soil, a living and dynamic environment, assumes a significant part in supporting plant development, controlling water stream, and cycling supplements. Regular farming practices, like monoculture and the exorbitant utilization of substance inputs, have been related with soil corruption. Unreasonable soil the executives adds to disintegration, loss of natural matter, and supplement consumption. Feasible horticultural works on, including cover trimming, crop pivot, and natural cultivating, rotate around supporting soil wellbeing. By cultivating different microbial networks, improving soil structure, and advancing supplement cycling, these practices add to the drawn out maintainability of rural frameworks.

Biodiversity preservation surfaces as one more basic thought in our investigation. Biodiversity is the texture of life that supports the versatility and flexibility of biological systems. Customary cultivating frameworks, portrayed by polyculture and blended trimming, innately support biodiversity by making heterogeneous scenes. Conversely, the broad reception of monoculture can prompt the rearrangements of scenes, making them more vulnerable to irritations, sicknesses, and ecological changes.

The investigation of the harmony between farming requirements and natural worries requires a comprehension of the job biodiversity plays in improving environment benefits and making vigorous, strong agrarian frameworks.

The mind boggling connection among farming and biodiversity is additionally highlighted by the crucial job of pollinators. Honey bees, butterflies, and different pollinators add to the generation of many blooming plants, including various harvest species. The biological system administration of fertilization is necessary to farming efficiency, and its downfall represents an immediate danger to food security. Impractical horticultural works on, including the utilization of specific pesticides and environment annihilation, imperil pollinator populaces. As we explore this equilibrium, the advancement of pollinator-accommodating cultivating rehearses and the security of regular natural surroundings become basic in supporting both horticulture and environmental wellbeing.

Water the executives arises as a focal fundamental in the investigation of the multifaceted connection among farming and environment. Horticulture is a significant shopper of water assets, and wasteful water system practices can prompt water shortage, especially in parched and semi-dry locales. Impractical water use adds to the consumption of springs and postures long haul difficulties for water supportability. Feasible water the board works on, including the reception of effective water system advances, water gathering, and the rebuilding of normal water biological systems, are pivotal in alleviating the ecological effect of horticulture and protecting water assets for people in the future.

The preservation of normal living spaces inside and around rural scenes assumes a vital part in keeping up with biological equilibrium. The development of horticultural wildernesses, frequently to the detriment of normal environments like timberlands, adds to living space misfortune and undermines biodiversity. Agroforestry, a methodology that incorporates trees and bushes into farming frameworks, gives a way to monitor biodiversity, further develop soil wellbeing, and improve biological system administrations. Riparian cradles and untamed life passages in horticultural scenes add to the safeguarding of regular natural surroundings, making interconnected networks that help assorted plant and creature species.

The hereditary variety of yields becomes the overwhelming focus in the investigation of this sensitive equilibrium. The dependence on a set number of high-yielding harvest assortments, an outcome of current modern farming, has prompted a decrease in crop hereditary variety. This absence of variety makes rural frameworks more powerless against arising bugs, infections, and changing ecological circumstances. Endeavors to advance and ration crop hereditary variety through drives, for example, seed banks, local area seed trades, and the development of customary and privately adjusted assortments become urgent in building strong and practical agrarian frameworks.

Coordinated bother the executives (IPM) arises as a reasonable way to deal with bug control that lines up with the investigation of this equilibrium. IPM tries to limit the utilization of synthetic pesticides while successfully overseeing nuisance populaces through a comprehensive comprehension of environmental elements. By integrating organic control techniques, like the utilization of normal hunters and parasites, social practices, and sensible compound mediations when fundamental, IPM epitomizes a methodology that keeps a harmony between rural efficiency and biological wellbeing.

The idea of regenerative horticulture adds a layer of goal to manageable practices by zeroing in on rural techniques that effectively add to the recovery of environments. Regenerative agribusiness looks to revamp soil wellbeing, upgrade biodiversity, and sequester carbon in soils. Practices, for example, agroforestry, cover editing, and rotational brushing structure basic parts of regenerative horticulture, underscoring a comprehensive methodology that tends to the reliance of farming and nature. As we investigate this equilibrium, regenerative horticulture arises as a reference point for those looking for groundbreaking practices that go past maintainability towards effectively reestablishing and upgrading biological systems.

Biological system benefits, the advantages that environments give to human social orders, possess a vital situation in the investigation of the harmony among horticulture and biology. These administrations incorporate fertilization of yields, water cleansing, soil richness, and environment guideline. Feasible rural practices that focus on environment wellbeing add to the proceeded with arrangement of these administrations. Preservation agribusiness, which limits soil unsettling influence and advances cover trimming, improves soil structure, diminishes disintegration, and supports environment administrations basic for horticulture.

The investigation of this fragile equilibrium likewise requires a thought of the social and monetary elements of farming. The livelihoods of millions of individuals rely upon agribusiness, especially in rustic areas of emerging nations. Manageable rural practices shouldn't just address biological worries yet in addition add to the prosperity of cultivating networks. This incorporates elevating fair admittance to land, assets, and markets, as well as encouraging strength notwithstanding environmental change and other outside challenges. The social component of maintainability requires the strengthening of ranchers, the consideration of different voices in dynamic cycles, and the making of rural frameworks that upgrade the personal satisfaction for those straightforwardly associated with food creation.

Chapter 6

Cultural and Social Dimensions

Social and social aspects assume a urgent part in forming the texture of social orders across the globe. These aspects incorporate a wide exhibit of components, going from convictions and values to customs, customs, and social organizations. Understanding and valuing these aspects is critical for encouraging amicability and collaboration inside assorted networks. In this investigation, we dig into the complex transaction of social and social aspects, analyzing their effect on people, networks, and the more extensive worldwide scene.

At the core of social aspects lie the well established convictions and values that characterize a general public. These key standards act as the structure blocks of social character, impacting people's insights, ways of behaving, and collaborations. Social variety, a sign of human social orders, mirrors the wealth and intricacy of these aspects. The mosaic of societies overall adds to a dynamic embroidery, cultivating culturally diverse trades and shared improvement.

Language remains as a foundation of social character, exemplifying the aggregate insight and articulations of a local area. Etymological variety reflects the fluctuated manners by which various social orders conceptualize and express their encounters. The subtleties implanted in dialects convey the pith of social legacy, forming the story of a group. In any case, etymological variety additionally presents difficulties, as viable correspondence becomes central in an interconnected world.

Past language, customs and customs comprise one more necessary part of social aspects. These practices act as soul changing experiences, stamping critical achieve-ments in people's lives. Whether it be birth services, weddings, or memorial services, customs give a feeling of congruity and having a place. Also, they go about as archives of social information, sending values and shrewdness starting with one age then onto the next.

Religion, a powerful power in molding social aspects, assumes a focal part in numerous social orders. It gives a structure to grasping the world, characterizing moral codes, and offering a feeling of direction. The exchange among religion and culture can be perplexing, as strict convictions frequently impact accepted practices, administration frameworks, and relational connections. This convergence shapes the shared perspective of a local area and adds to the development of cultural designs.

Social establishments, like family, training, and administration, are crucial parts of the social aspects that support human social orders. The nuclear family, as a microcosm of society, confers values, standards, and socialization to people. It fills in as the essential wellspring of daily reassurance and personality arrangement, establishing the groundwork for one's position in the more extensive social texture.

Training, as a cultural establishment, assumes a significant part in molding people's viewpoints and encouraging social union. Schools and school systems reflect social qualities and needs, impacting the transmission of information and the improvement of decisive reasoning abilities. Besides, training goes about as an extension between ages, working with the exchange of social legacy and the development of a common cultural ethos.

Administration structures, whether popularity based, dictator, or some in the middle between, are characteristic for the social components of any general public. How power is conveyed, and choices are made mirrors the fundamental qualities and needs of a local area. Successful administration is pivotal for keeping social control, guaranteeing equity, and tending to the requirements of different populaces. The connection among administration and culture is dynamic, with each impacting and molding the other in a continuous exchange.

As social orders develop, the effect of globalization turns out to be progressively articulated, impacting social and social aspects on a worldwide scale. The interconnectedness worked with by innovation, exchange, and correspondence has prompted the mixing and dispersion of societies. While this globalization has achieved open doors for cooperation and understanding, it has additionally raised worries about social homogenization and the disintegration of unmistakable social personalities.

Social appointment, a peculiarity exacerbated by globalization, features the intricacies of social trade. The acquiring or impersonation of components from one culture by another can prompt appreciation and advancement. Nonetheless, when done unfeelingly or without affirmation, it can sustain generalizations and add to social commodification. Finding some kind of harmony between praising variety and keeping away from social allocation is a fragile undertaking that requires social responsiveness and shared regard.

In the advanced age, online entertainment stages have become strong specialists of social trade and impact. These stages interface people across topographical limits, empowering the sharing of thoughts, values, and social articulations. Nonetheless, they likewise present difficulties, as the fast dispersal of data can prompt the spread of falsehood, generalizations, and the enhancement of troublesome stories. Exploring the computerized scene requires a nuanced comprehension of its effect on social and social aspects.

The idea of social relativism stresses the significance of grasping social practices inside their own unique situation, without forcing outside decisions. Perceiving that various social orders have extraordinary chronicles, values, and standards is fundamental for cultivating culturally diverse comprehension. Social relativism challenges ethnocentrism, empowering people to move toward new social practices with a receptive outlook and an eagerness to participate in significant exchange.

In the domain of social aspects, the idea of civil rights becomes the overwhelming focus. Civil rights advocates for the fair conveyance of assets, open doors, and privileges inside a general public. It points out foundational disparities in light of elements like race, orientation, financial status, and different components of character. Addressing civil rights requires aggregate endeavors to destroy prejudicial designs and establish comprehensive conditions that engage all people.

Multifacetedness, a structure that thinks about the interconnected idea of social characters, upgrades how we might interpret social aspects. It perceives that people experience numerous layers of honor and persecution in view of their meeting personalities. By embracing a multifaceted point of view, social orders can foster more comprehensive arrangements and practices that address the remarkable difficulties looked by people with assorted foundations and characters.

The job of media in forming social and social aspects couldn't possibly be more significant. News sources, including news associations, film, TV, and web-based entertainment stages, impact general assessment and add to the development of cultural accounts. Portrayal in media matters, as it shapes view of various gatherings and impacts cultural mentalities. Different and comprehensive media depictions can challenge generalizations and add to a more impartial and figuring out society.

Character, a perplexing exchange of social, social, and individual aspects, is at the center of individual and aggregate encounters. Individual personality includes factors like orientation, nationality, sexual direction, and the sky is the limit from there, while aggregate character is formed by shared social encounters and authentic accounts. Investigating and understanding personality is a continuous interaction that includes self-reflection, exchange, and an acknowledgment of the liquid and dynamic nature of character builds.

With regards to social and social aspects, relocation arises as a critical peculiarity that shapes the variety and elements of social orders. Relocation unites people from various social foundations, making multicultural networks. While relocation can prompt social improvement and the trading of thoughts, it additionally presents difficulties connected with joining, segregation, and the safeguarding of social legacy.

The powerful idea of social and social aspects is obvious in the advancement of cultural standards and values over the long haul. Social advancement frequently includes testing existing standards and supporting for change to address fundamental shameful acts. Developments for social equality, orientation equity, LGBTQ+ privileges, and other civil rights purposes show the extraordinary force of aggregate activity in reshaping social and social scenes.

The effect of social and social aspects on psychological well-being is a basic area of investigation. Social elements shape people's view of psychological well-being, impacting help-chasing ways of behaving and the disgrace related with psychological well-being issues. Furthermore, social encouraging groups of people and local area perspectives assume a critical part in mental prosperity. Figuring out the convergence of culture, social elements, and emotional wellness is fundamental for growing socially skilled and comprehensive psychological well-being rehearses.

Natural supportability, however frequently viewed as a biological issue, is interlaced with social and social aspects. Native societies, specifically, have longstanding associations with the climate, seeing it as necessary to their character and prosperity. The abuse of normal assets, environmental change, and natural debasement lopsidedly influence weak networks, highlighting the interconnectedness of biological and social frameworks.

6.1 Insight into the cultural significance of lift irrigation in various communities

Lift water system, a training profoundly implanted in the farming practices of different networks, holds significant social importance. This water system strategy includes lifting water from a lower to a higher height, empowering the development of yields in regions where regular water sources may not be promptly accessible. Past its viable utility, lift water system is entwined with social accounts, ceremonies, and mutual personality, mirroring the cooperative connection among networks and their agrarian practices.

In numerous agrarian social orders, the land isn't only a method for creation however a storehouse of social legacy and collective personality. The development of yields, worked with by water system strategies like lift water system, turns into a holy endeavor, interfacing people to the patterns of nature and the rhythms of life. The social meaning of lift water system is much of the time communicated through customs and functions that mark the start of the establishing season, the appearance of storms, or the reaping of yields.

In India, for instance, the act of lift water system has profound roots in the social texture of country networks. The notorious stepwells, known as 'baolis' or 'vavs,' embody customary water-lifting structures that filled both reasonable and emblematic needs. These stepwells, frequently complicatedly planned with engineering components, became public spaces where individuals assembled not exclusively to bring water yet in addition to take part in friendly communications and social exercises. The demonstration of drawing water from these stepwells was not simply a utilitarian errand but rather a ceremonial practice implanted in the regular routines of networks.

Additionally, in pieces of Africa, where farming supports livelihoods and networks, lift water system rehearses are pervaded with social importance. The common exertion expected for overseeing water system channels, the festival of effective harvests, and the sharing of water assets all add to a common social story. At times, old stories and oral customs pass down the insight of proficient water the board, stressing the significance of lift water system for food as well as a social legacy to be saved and given to people in the future.

The emblematic worth of water in different strict and social customs further enhances the social meaning of lift water system. Water, frequently seen as a nurturing force, is vital to numerous strict services and customs. Lift water system frameworks, by guaranteeing a steady water supply for farming purposes, line up with the social worship for water as a wellspring of life, fruitfulness, and flourishing. The demonstration of lifting water turns into a substantial articulation of the local area's association with the land and its reliance on regular components for food.

Notwithstanding its job in social ceremonies, lift water system shapes the social elements of networks. The helpful administration of water system framework cultivates a feeling of divided liability and relationship between local area individuals. Water dissemination, not set in stone through aggregate dynamic cycles, mirror the common agreement overseeing the impartial utilization of a typical asset. This cooperative methodology guarantees the effective use of water as well as builds up friendly securities inside the local area.

The social meaning of lift water system is likewise apparent in the legends and narrating customs of networks that training this strategy. Stories of legendary creatures or divinities related with water, water system, and agribusiness have large amounts of these accounts. These accounts serve as diversion as well as for the purpose of communicating social qualities, environmental insight, and the significance of manageable practices. Through narrating, networks support the interconnectedness of their social personality with the common habitat and the rural practices that support them.

Besides, the flexibility of lift water system strategies to assorted geological and climatic circumstances highlights its social versatility. Various people group have created extraordinary procedures and innovations fit to their particular scenes, mirroring a profound comprehension of the nearby biology. This information is much of the time elapsed down through ages, framing a vital piece of the social legacy that shapes the local area's relationship with the land.

The social meaning of lift water system isn't static; it advances because of changes in financial, natural, and mechanical elements. As people group wrestle with the effects of environmental change, moving atmospheric conditions, and the requirement for supportable rural practices, the social stories around lift water system might go through changes. Now and again, networks might coordinate present day innovation with conventional works on, making half and half frameworks that offset social legacy with contemporary necessities.

The conservation of social importance in lift water system rehearses faces difficulties despite urbanization, industrialization, and globalization. As conventional agrarian social orders go through quick change, there is a gamble of social disintegration as more youthful ages create some distance from farming occupations. The shift towards motorized and modern cultivating may likewise underestimate conventional works on, lessening the social associations with lift water system.

Endeavors to record, celebrate, and revive the social meaning of lift water system are significant for keeping up with the strength of these practices. Social foundations, instructive projects, and local area drove drives assume a fundamental part in saving and advancing the rich embroidery of customs, stories, and customs related with lift water system. Underlining the social significance close by the functional utility of these strategies can add to a comprehensive comprehension that encourages reasonable practices established in social legacy.

All in all, lift water system rises above its utilitarian reason to turn into a social standard for some networks all over the planet. The demonstration of lifting water isn't just a mechanical interaction however a custom, a common undertaking, and a connection to social personalities molded by the land. The imagery of water, the helpful administration of assets, and the entwining of rural practices with social accounts highlight the profundity of the association between lift water system and the social legacy of different networks. As social orders explore the intricacies of modernization and ecological difficulties, perceiving and protecting the social meaning of lift water system becomes basic for cultivating practical, strong, and socially rich agrarian scenes.

6.2 Examination of the impact on social structures and rural livelihoods

The assessment of the effect on friendly designs and country jobs uncovers the unpredictable interaction between different variables that shape the texture of rustic networks. Social designs, frequently well established in custom and local area elements, go through changes affected by monetary, natural, and mechanical changes. All the while, provincial jobs, which are unpredictably attached to agribusiness and regular assets, experience moves that can have sweeping ramifications for the prosperity of people and networks.

In numerous country social orders, social designs are firmly entwined with agrarian works on, framing the foundation of local area life. The agrarian lifestyle frequently includes aggregate independent direction, shared liabilities, and a solid feeling of local area. In any case, the effect of outside powers, like globalization and urbanization, can disturb these customary social designs. The acquaintance of market-driven approaches with farming might prompt changes in land possession designs, adjusting power elements inside networks.

The shift towards cash harvests and business cultivating, driven by market requests, can significantly affect provincial social designs. Customary resource cultivating, where families develop crops essentially for their own utilization, may give way to a more market-situated approach. This progress can prompt changes in land use, as ranchers allot more land to cash crops, possibly influencing food security and customary cultivating rehearses.

Landownership designs assume a critical part in molding social designs in rustic networks. In numerous social orders, land isn't simply a financial resource however a wellspring of character and economic wellbeing. Changes in land dispersion, whether through market influences, government strategies, or legacy rehearses, can prompt changes in power elements and impact the social progressive system inside country settings. The effect is many times felt at the grassroots level, where people and families explore these progressions as they continued looking for practical livelihoods.

The coming of innovation, including automated cultivating gear and accuracy agribusiness, further adds to shifts in rustic social designs. While innovation can improve efficiency and productivity, it likewise requires particular abilities, possibly prompting a division of work inside networks. This division might influence customary orientation jobs, with specific errands turning out to be more motorized and others requiring an alternate range of abilities. The incorporation of innovation can set out new monetary open doors yet may likewise extend existing financial incongruities inside rustic populaces.

Schooling, a critical determinant of social versatility, assumes a vital part in forming country social designs. Admittance to quality instruction can enable people, opening up open doors past conventional horticultural occupations. In any case, difficulties like deficient framework, restricted assets, and social standards might obstruct instructive fulfillment in provincial regions. The effect of instruction on friendly designs reaches out past individual strengthening to local area improvement, as taught people add to the general prosperity and flexibility of rustic social orders.

Relocation, driven by monetary open doors or natural elements, is another component that impacts provincial social designs. The outmigration of youthful grown-ups looking for business can prompt segment lopsided characteristics inside country networks. The takeoff of a critical part of the working-age populace might influence the social texture, as networks wrestle with issues like maturing populaces, less accessible workers for farming, and changes in local area elements.

The effect of movement on rustic jobs is complex. While settlements from travelers can add to the monetary prosperity of provincial families, the shortfall of relatives can strain social ties and disturb customary emotionally supportive networks. Besides, the arrival of travelers, assuming they decide to return, may bring groundbreaking thoughts and points of view, adding to both positive and testing changes in country social designs.

Ecological variables, including environmental change and normal asset corruption, present huge difficulties to provincial occupations and social designs. Changes in atmospheric conditions, like delayed dry seasons or eccentric precipitation, can straightforwardly affect horticultural efficiency. The subsequent vulnerabilities might prompt changes in customary occupation methodologies as networks adjust to new climatic real factors. Moreover, rivalry for scant regular assets can heighten, possibly prompting clashes inside and between rustic networks.

The flexibility of rustic social designs is firmly connected to the capacity of networks to adjust to natural difficulties. Maintainable farming practices, local area based normal asset the executives, and environment tough procedures are fundamental for shielding both social designs and provincial jobs. Cooperation between networks, state run administrations, and non-legislative associations becomes basic to address the mind boggling and interconnected difficulties presented by natural changes.

Government strategies and mediations assume a vital part in molding the effect on country social designs and jobs. Agrarian changes, land residency strategies, and country advancement drives can possibly either reinforce or sabotage existing social designs. Comprehensive and participatory policymaking that considers the assorted requirements and points of view of provincial networks is fundamental for encouraging manageable turn of events.

The social elements of country life, profoundly interlaced with social designs and livelihoods, add to the uniqueness of each provincial local area. Social legacy, customs, and neighborhood information shape the manner in which people see and collaborate with their current circumstance. Protecting and advancing social variety inside country regions isn't just significant for the prosperity of networks yet in addition for keeping a rich embroidery of human encounters and points of view.

The effect on friendly designs and rustic jobs isn't uniform across locales or networks. Setting explicit variables, for example, nearby administration structures, social standards, and verifiable heritages, add to the variety of encounters inside rustic settings. Perceiving this variety is essential for creating designated mediations that address the one of a kind difficulties and valuable open doors looked by changed networks.

Taking everything into account, the assessment of the effect on friendly designs and rustic vocations highlights the intricacy and interconnectedness of elements that shape provincial networks. Social designs, well established in custom and local area elements, go through changes impacted by monetary, ecological, and mechanical changes. The flexibility of country social designs is firmly connected to the versatility of networks to difficulties, for example, changes in land use, mechanical headways, training, movement, ecological elements, and government approaches.

Safeguarding the social components of country life is fundamental for keeping up with the novel character of every local area. As rustic regions explore the powerful powers of progress, cultivating comprehensive and feasible improvement becomes basic. Enabling provincial networks through training, supporting manageable horticultural works on, tending to natural difficulties, and guaranteeing comprehensive administration are necessary parts of building strong social designs and livelihoods inside rustic scenes. The assessment of these elements gives significant bits of knowledge to policymakers, scientists, and specialists pursuing the comprehensive advancement of country networks.

6.3 Illustrating the integration of lift irrigation into local customs

The coordination of lift water system into nearby traditions embodies the amicable concurrence of customary practices and mechanical development inside different social scenes. Lift water system, a technique for moving water to higher heights for farming purposes, has consistently woven itself into the texture of different networks, enhancing nearby traditions and supporting rural customs. This joining isn't simply a specialized transformation however a unique interaction where the useful utility of lift water system combines with the social qualities, customs, and local area elements that characterize a specific region.

In numerous districts where farming is a lifestyle, the presentation of lift water system advances has turned into an extraordinary power. This mix is many times driven by the squeezing need to improve rural efficiency, particularly in regions confronting water shortage or unpredictable water appropriation. As opposed to being seen as a takeoff from conventional practices, lift water system turns into an essential expansion that supplements and fortifies existing horticultural traditions.

Neighborhood customs connected with water the board and agribusiness are well established in the social history of networks. Practices, for example, mutual water-sharing plans, customary water system ceremonies, and the festival of farming celebrations are basic parts of these traditions. The joining of lift water system flawlessly lines up with these traditions, offering a cutting edge answer for age-old difficulties while regarding and saving the social importance connected to water and farming.

One critical illustration of the coordination of lift water system into neighborhood customs can be seen in the terraced fields of the Himalayan district. The conventional horticultural practices in these difficult territories include developing yields on painstakingly built porches to augment water use and forestall soil disintegration. With the presentation of lift water system, these terraced scenes have adjusted to incorporate current siphoning frameworks that effectively transport water to more significant levels. This combination not just supports the rural legacy of the area yet additionally improves the flexibility of networks confronting the effects of environmental change.

The mutual parts of lift water system adjust intimately with the aggregate idea of numerous conventional traditions in agrarian social orders. Local area oversaw water system frameworks, where ranchers share the obligation of keeping up with and working lift water system foundation, build up a feeling of aggregate proprietorship. This cooperative methodology guarantees the impartial conveyance of water as well as cultivates social attachment inside the local area.

Customs and services related with water and agribusiness take on new aspects with the reconciliation of lift water system. In certain societies, the demonstration of lifting water turns into an emblematic motion during strict or occasional functions. The musical sound of water being lifted and conveyed through the fields might be joined by customary melodies, moves, or supplications, making a combination of old traditions and current innovation.

In specific locales, the planning of lift water system activities lines up with divine occasions or social schedules. For example, the beginning of the rainstorm season, a vital period for farming in numerous tropical locales, may match with shared exercises connected with lift water system. This synchronization mirrors the profound association between farming practices, ecological cycles, and social customs, stressing the relationship of the local area with the normal world.

The coordination of lift water system additionally meets with orientation elements inside numerous conventional social orders. In certain societies, certain horticultural undertakings, including water the executives, have generally been related with explicit orientation jobs. The presentation of lift water system innovations can challenge these orientation standards by making specific undertakings more open and less work concentrated. This change improves the productivity of horticultural practices as well as adds to more evenhanded orientation relations inside country networks.

Neighborhood information, one more essential part of social traditions, assumes a critical part in the effective reconciliation of lift water system. Native insight about soil types, crop assortments, and water dissemination designs illuminates the transformation regarding lift water system frameworks to suit the particular requirements of a locale. This cooperative trade between conventional information and present day innovation brings about creative arrangements that are established in the nearby setting.

With regards to lift water system, the maintainability of water assets turns into a common social concern. Networks that incorporate lift water system into their traditions frequently foster guidelines and standards for capable water use. Customary practices, for example, water revolution plans, occasional water-sharing arrangements, and local area drove water protection drives line up with the reasonable utilization of lift water system advancements. This combination supports that innovative headways can coincide with social qualities to address ecological difficulties.

The imagery of water, respected in numerous social practices, takes on added importance with the coordination of lift water system. Water, frequently seen as a nurturing force, turns into an unmistakable connection between the past and the present. The demonstration of lifting water turns into a formal work on, interfacing people to the immortal pattern of development, gather, and reestablishment. The combination of social imagery with the commonsense need of water transport through lift water system changes a utilitarian undertaking into a hallowed demonstration.

In certain areas, the presence of lift water system structures becomes tourist spots with social importance. Water towers, siphoning stations, or other framework related with lift water system might get emblematic worth inside the local area. These designs, while filling a functional need, likewise become visual portrayals of the local area's capacity to adjust to changing conditions while saving its social personality.

The coordination of lift water system into nearby traditions stretches out past the agrarian field to influence different parts of local area life. Monetary exercises, social designs, and, surprisingly, creative articulations might go through changes as networks embrace the potential outcomes presented by further developed water access. For instance, expanded rural efficiency worked with by lift water system can invigorate financial broadening, prompting the improvement of neighborhood markets, distinctive artworks, and other non-horticultural livelihoods.

Social designs, frequently impacted by horticultural practices, adjust to the progressions achieved by lift water system. The helpful administration of water assets cultivates a feeling of shared liability and common reliance. Conventional pecking orders might advance as local area individuals on the whole take part in dynamic cycles connected with the activity and upkeep of lift water system frameworks. This co-operative methodology guarantees the maintainability of water use as well as fortifies social securities inside the local area.

The coordination of lift water system into neighborhood customs isn't without challenges. The presentation of new advancements might confront opposition from networks familiar with conventional practices. Social dormancy, financial limitations, or worries about the natural effect of mechanical intercessions can present obstructions to reception. Powerful mix requires a nuanced comprehension of neighborhood customs, open exchange, and local area contribution in dynamic cycles.

Government strategies and outer mediations likewise assume a critical part in molding the combination of lift water system into neighborhood customs. Strong arrangements that perceive and esteem native information, advance feasible water use, and guarantee impartial admittance to lift water system innovations add to fruitful coordination. Interestingly, hierarchical methodologies that ignore social subtleties or focus on momentary additions over long haul manageability might confront opposition and unseen side-effects.

All in all, the coordination of lift water system into neighborhood customs fills in as a strong delineation of the unique transaction among custom and development. Instead of review mechanical headways as problematic powers, networks that effectively coordinate lift water system embrace an all encompassing methodology that regards social qualities, natural contemplations, and the pragmatic necessities of horticulture. This incorporation is a demonstration of the versatile limit of networks, where old traditions and current innovations merge to make feasible and socially resounding arrangements.

As social orders keep on exploring the difficulties of water shortage, environmental change, and advancing farming practices, the case of incorporating lift water system into nearby traditions gives significant bits of knowledge into the potential for conjunction among custom and progress.

The mix of lift water system into neighborhood customs addresses an intriguing intersection of old practices and current mechanical headways, winding around together the functional necessities of horticulture with the well established social acts of different networks. Lift water system, the most common way of raising water from lower to higher rises for horticultural purposes, has become in excess of a simple mechanical mediation; it has turned into a residing demonstration of the versatile idea of social orders, where development fits with customary traditions, molding the farming scenes and social personalities of different districts.

In numerous provincial social orders, farming isn't simply a financial movement; it is a lifestyle profoundly laced with social practices and mutual bonds. The presentation of lift water system innovations consistently coordinates with these customs, offering an answer for age-old difficulties while protecting the social qualities connected to water the executives and cultivating. As opposed to a troublesome power, lift water system turns into a device that upgrades and supports conventional horticultural traditions, framing a many-sided embroidery that connects the past, present, and eventual fate of networks.

One illustrative illustration of the mix of lift water system into nearby traditions can be found in the rice patios of Southeast Asia, especially in areas like the Philippine Cordilleras. These terraced scenes, cunningly planned by native networks over hundreds of years, are a demonstration of the multifaceted information on water stream and land the board. With the approach of lift water system advances, these networks have adjusted their customary practices to consolidate present day siphoning frameworks, keeping up with the manageability and social meaning of their rice patios.

The public parts of lift water system adjust consistently with the aggregate idea of numerous conventional traditions in agrarian social orders. Local area oversaw water system frameworks, where ranchers work together in the upkeep and activity of lift water system foundation, support a feeling of shared possession and obligation. This cooperative methodology goes past guaranteeing the fair dissemination of water; it cultivates social union inside the local area, reinforcing the bonds that integrate people in their common quest for food and prosperity.

Customs and functions related with water and farming interpretation of new aspects with the mix of lift water system. In certain societies, the demonstration of lifting water turns into an emblematic motion during strict or occasional services.

The musical sound of water being lifted and dispersed through the fields might be joined by customary tunes, moves, or petitions, making a combination of old traditions and current innovation. This mixing of the holy and the down to earth mirrors the interconnectedness of farming with the profound and social components of local area life.

The planning of lift water system activities frequently lines up with divine occasions or social schedules, building up the profound association between rural practices, ecological cycles, and social customs. For instance, the inception of lift water system exercises might concur with the beginning of the storm season, an essential period for the majority agrarian social orders. This synchronization accentuates the complex exchange between the items of common sense of water the board and the social customs that mark the farming schedule.

In specific locales, the presentation of lift water system has impacted orientation elements inside conventional social orders. Rural undertakings, including water the board, have generally been related with explicit orientation jobs. The combination of lift water system advancements can challenge these orientation standards by making specific undertakings more open and less work concentrated. This change improves the proficiency of farming practices as well as adds to more evenhanded orientation relations inside country networks, cultivating a feeling of consideration and shared liability.

The maintainability of water assets turns into a common social worry with the coordination of lift water system. Networks frequently foster standards and standards for capable water use, drawing from customary practices, for example, water turn plans, occasional water-sharing arrangements, and local area drove water protection drives. This combination supports that innovative headways can coincide with social qualities to address ecological difficulties, cultivating a feeling of natural stewardship inside the local area.

The imagery of water, respected in numerous social customs, takes on added importance with the reconciliation of lift water system. Water, frequently seen as a nurturing force, turns into an unmistakable connection between the past and the present. The demonstration of lifting water turns into a formal work on, interfacing people to the immortal pattern of development, collect, and recharging. The combination of social imagery with the viable need of water transport through lift water system changes a utilitarian errand into a consecrated demonstration, making a story that rises above ages.

In certain locales, the presence of lift water system structures becomes tourist spots with social importance. Water towers, siphoning stations, or other framework related with lift water system might gain representative worth inside the local area. These designs, while filling a useful need, likewise become visual portrayals of the local area's capacity to adjust to changing conditions while protecting its social character. They stand as demonstrations of the versatility of customary traditions despite innovative advancement.

The reconciliation of lift water system into nearby traditions reaches out past the farming field to influence different parts of local area life. Monetary exercises, social designs, and, surprisingly, imaginative articulations might go through changes as networks embrace the potential outcomes presented by further developed water access. Expanded farming efficiency worked with by lift water system can animate monetary expansion, prompting the advancement of nearby business sectors, distinctive specialties, and other non-agrarian livelihoods. This monetary broadening adds to the energy and manageability of provincial networks, making a harmony among custom and progress.

Social designs, frequently impacted by agrarian practices, adjust to the progressions achieved by lift water system. The helpful administration of water assets cultivates a feeling of shared liability and common reliance. Customary orders might advance as local area individuals altogether participate in dynamic cycles connected with the activity and upkeep of lift water system frameworks. This cooperative methodology guarantees the manageability of water use as well as reinforces social securities inside the local area, encouraging a feeling of solidarity and aggregate reason.

The incorporation of lift water system into neighborhood customs isn't without its difficulties. The presentation of new innovations might confront obstruction from networks familiar with customary practices. Social latency, financial limitations, or worries about the ecological effect of innovative mediations can present obstructions to reception. Viable joining requires a nuanced comprehension of neighborhood customs, open exchange, and local area contribution in dynamic cycles. It requires a cooperative methodology where innovative intercessions line up with the social qualities and needs of the local area.

Government strategies and outer mediations likewise assume a critical part in forming the coordination of lift water system into nearby traditions. Steady approaches that perceive and esteem native information, advance practical water use, and guarantee fair admittance to lift water system innovations add to effective reconciliation. Conversely, hierarchical methodologies that ignore social subtleties or focus on momentary increases over long haul manageability might confront opposition and unseen side-effects.

The safeguarding and festivity of social variety inside the setting of lift water system are fundamental for keeping up with the special character of every local area. As social orders keep on exploring the intricacies of water shortage, environmental change, and advancing rural practices, the joining of lift water system into nearby traditions gives significant experiences into the potential for conjunction among custom and progress. It fills in as a reference point for manageable improvement that regards social legacy, advances local area prosperity, and embraces the cooperative energy between age-old traditions and contemporary developments.

Chapter 7

Global Perspectives

In the steadily advancing embroidery of our reality, worldwide viewpoints have turned into the focal point through which we comprehend and explore the intricacies of an interconnected and reliant planet. The twenty-first century has seen an exceptional speed increase in globalization, reshaping social orders, economies, and societies across mainlands. This peculiarity isn't only a theoretical idea however a living power that pervades each part of our lives, impacting political choices, monetary strategies, social trades, and innovative headways.

One of the central qualities of the contemporary time is the strengthening of worldwide interconnectivity. Propels in correspondence innovations, like the web and online entertainment, have changed the world into a huge organization where data streams quickly and limits obscure. Geographic distances have become less weighty as people, organizations, and legislatures take part progressively communications, rising above borders and reshaping the customary ideas of reality.

In this interconnected world, occasions in a single corner of the globe swell across landmasses, influencing lives and molding predeterminations. A monetary emergency in one nation can set off a cascading type of influence, influencing economies around the world.

Catastrophic events, once bound to territorial outcomes, presently draw in worldwide consideration as the world observers their prompt and broad effects. The Coronavirus pandemic fills in as an obvious sign of the interconnectedness of our worldwide local area, featuring the common weaknesses and the basic for cooperative reactions to transnational difficulties.

Monetary globalization has been a main thrust behind the reshaping of worldwide points of view. The reconciliation of public economies into a solitary, reliant framework has prompted the phenomenal development of products, administrations, capital, and work across borders. Worldwide organizations, with tasks spreading over different nations, assume a critical part in molding the worldwide monetary scene. The ascent of worldwide stockpile chains has made creation processes more productive yet additionally more defenseless to disturbances, as exhibited by the difficulties presented by the pandemic.

The elements of worldwide exchange have moved, with arising economies stating their presence and customary monetary forces to be reckoned with confronting new difficulties. The ascent of China as a worldwide financial force to be reckoned with has reshaped the international scene, testing the strength of Western countries and inciting a reconsideration of worldwide power structures. Economic deals, like the Transoceanic Association (TPP) and the Local Complete Monetary Organization (RCEP), mirror the developing financial coalitions that rise above customary international limits.

Globalization has changed monetary scenes as well as introduced social trades on an uncommon scale. The crash of different societies has prompted a rich embroidery of worldwide impacts, molding everything from cooking and design to music and workmanship. The social homogenization dreaded by certain has given way to a lively mosaic of variety, where thoughts, customs, and points of view mix, making a world-wide culture that is both dynamic and diverse.

In any case, this social trade isn't without challenges. The fast scattering of data and thoughts through advanced stages has prompted worries about social government, where predominant societies eclipse and minimize more modest, less persuasive ones. The conservation of social personality despite globalization is a steady battle for some networks, provoking conversations about the requirement for social responsiveness and the security of social legacy in the worldwide field.

Political scenes, as well, have been reshaped by the powers of globalization. Public states wrestle with the need to offset homegrown interests with the requests of an interconnected world. The ascent of egalitarian developments in different pieces of the globe mirrors a reaction against saw dangers to public personality and power. Worldwide difficulties, for example, environmental change, movement, and general wellbeing emergencies require cooperative arrangements, frequently testing the constraints of global organizations and political relations.

The idea of public safety has extended to incorporate conventional military dangers as well as forward thinking difficulties, for example, digital fighting, financial weaknesses, and worldwide pandemics. The interconnected idea of current cultures implies that a disturbance in one region can have flowing impacts, making the conventional thoughts of safety progressively lacking in tending to the mind boggling and inter-related difficulties of the twenty-first 100 years.

Ecological issues, especially the approaching ghost of environmental change, high-light the genuinely worldwide nature of the difficulties we face. The effect of human exercises in the world rises above public lines, and deliberate global endeavors are fundamental to moderate the results. The Paris Understanding, a milestone accord endorsed by countries all over the planet, addresses an aggregate obligation to tending to environmental change and highlights the requirement for worldwide participation despite existential dangers.

The computerized upset plays had a focal impact in molding worldwide viewpoints, offering the two open doors and difficulties. The democratization of data through the web has engaged people, empowering them to get to information, associate with others, and take part out in the open talk on an uncommon scale. Web-based entertainment stages have become incredible assets for activism, empowering grassroots developments to prepare and impact change.

Nonetheless, the computerized time likewise delivers new difficulties, including worries about security, the spread of deception, and the effect of innovation on work. The ascent of man-made reasoning (simulated intelligence) and mechanization can possibly change ventures and work markets, bringing up issues about the fate of work and the requirement for new ideal models of schooling and abilities improvement to adjust to the advancing position scene.

Worldwide points of view reach out past the domains of legislative issues, financial aspects, and innovation to envelop the domains of morals and common liberties. The interconnected idea of the world requests a reconsideration of moral structures that rise above public limits. Basic freedoms infringement in one region of the planet are not generally confined concerns; they reverberate universally, provoking calls for worldwide mediation and responsibility.

The worldwide local area faces moral problems in exploring the moral ramifications of arising advances, like biotechnology and hereditary designing. Inquiries regarding the moral utilization of information, the effect of man-made intelligence on security, and the possible abuse of cutting edge innovations challenge social orders to lay out moral rules that reflect shared values and standards.

Schooling, as well, goes through a change in the time of globalization. The requirement for a worldwide viewpoint is progressively perceived in instructive educational programs, planning understudies to explore the intricacies of an interconnected world. Internationalization of schooling, trade projects, and diverse encounters become fundamental parts of cultivating a worldwide mentality.

As we explore the difficulties and amazing open doors introduced by worldwide points of view, the job of global foundations becomes urgent. Associations like the Assembled Countries (UN), the World Exchange Association (WTO), and the World Wellbeing Association (WHO) assume critical parts in tending to worldwide difficulties and encouraging participation among countries. Be that as it may, these organizations likewise face analysis for their apparent failures and the difficulties of adjusting to a quickly influencing world.

The mission for worldwide viewpoints requires a pledge to discretion and exchange as fundamental devices for settling clashes and building spans between countries. The specialty of tact turns out to be significantly more basic in a period where the results of political choices rise above public boundaries. The difficulties of atomic multiplication, territorial struggles, and the outcast emergency request political arrangements that focus on joint effort over conflict.

The job of non-legislative associations (NGOs) and common society couldn't possibly be more significant in forming worldwide viewpoints. Promotion gatherings, compassionate associations, and grassroots developments assume urgent parts in considering legislatures and partnerships responsible, enhancing minimized voices, and driving social change. The force of aggregate activity is apparent in developments resolving issues, for example, environmental change, common freedoms, and civil rights.

The groundbreaking idea of worldwide points of view stretches out to the domain of medical care. The Coronavirus pandemic fills in as a distinct sign of the interconnectedness of worldwide wellbeing. Infections perceive no boundaries, and the spread of irresistible sicknesses represents a common danger that requires worldwide collaboration in exploration, counteraction, and reaction. The pandemic highlights the need to fortify worldwide wellbeing framework and address wellbeing variations on a worldwide scale.

All in all, worldwide viewpoints are not simply a hypothetical structure but rather a lived reality that shapes our regular routines, our goals, and our common future. The difficulties and chances of an interconnected world constrain us to rise above thin limits and develop an outlook that embraces the extravagance of variety while perceiving our common mankind. In exploring the intricacies of the twenty-first hundred years, the capacity to embrace worldwide viewpoints becomes a scholarly activity as well as a key expertise for people, networks, and countries the same.

7.1 Comparative analysis of lift irrigation practices in different regions

The act of lift water system fills in as a basic part of farming water the executives, assuming a urgent part in giving water to crops in locales where normal water sources are not effectively open or adequate. A relative examination of lift water system rehearses across various districts offers important experiences into the shifted approaches, difficulties, and developments related with this fundamental farming strategy.

One of the essential qualifications in lift water system rehearses lies in the geological and geographical qualities of the areas where it is utilized. In rocky or uneven territories, where gravity-took care of water system is frequently unreasonable, lift water system turns into a fundamental arrangement. The use of siphons or other lifting components becomes important to ship water from lower to higher rises, guaranteeing that horticultural terrains arranged uphill get the necessary water system.

In the fields, where regular angles might work with gravity-based water system frameworks, lift water system is as yet utilized in unambiguous circumstances. Factors, for example, lopsided appropriation of water sources, varieties in land geology, and the need to draw water from more profound sources can require the utilization of lift water system advances, even in somewhat level locales.

The innovation utilized in lift water system frameworks shows huge varieties in light of neighborhood necessities, asset accessibility, and monetary contemplations. In certain districts, diesel or electric siphons are normally utilized for lifting water, while in others, imaginative innovations, for example, sun based controlled siphons are acquiring conspicuousness, driven by a developing accentuation on maintainable and energy-productive horticultural practices.

The decision of lifting instrument is much of the time affected by the accessibility and cost of fuel sources. In locales with dependable admittance to power, electric siphons might be the favored decision. In any case, in regions where power is scant or temperamental, diesel-controlled siphons or elective energy sources like sunlight based and wind become appealing choices. The similar investigation of lift water system rehearses accordingly mirrors the convergence of mechanical development, monetary contemplations, and nearby asset accessibility.

Water source the executives is a basic part of lift water system, and the relative examination uncovers different methodologies in various districts. A few regions depend on surface water bodies like streams, lakes, or repositories as the essential hotspot for lift water system, while others tap into groundwater through wells or boreholes. The supportability of lift water system rehearses is intently attached to the administration and preservation of these water sources, with over-extraction representing a huge danger to both the climate and the drawn out suitability of the horticultural frameworks.

The ecological effect of lift water system rehearses is a significant thought in the near examination. The extraction of water from normal sources, particularly in locales previously wrestling with water shortage, can prompt natural irregular characteristics, soil debasement, and other ecological worries. Manageable lift water system rehearses consolidate measures to limit natural effect, for example, the utilization of water-saving advances, re-energizing springs, and taking on soil preservation rehearses.

Social and social factors additionally shape the near scene of lift water system rehearses. Customary cultivating networks might have laid out one of a kind strategies for lift water system that are well established in nearby information and practices. The safeguarding of these customary strategies, frequently went down through ages, adds to the variety of lift water system techniques across various areas. Moreover, people group based ways to deal with water the executives, where ranchers by and large put resources into and oversee lift water system frameworks, are common in specific districts, cultivating a feeling of shared liability and collaboration.

Financial contemplations assume a focal part in forming lift water system works on, impacting the decision of innovation, water source the executives techniques, and generally framework plan. The reasonableness of lifting systems, functional expenses, and the profit from speculation for ranchers are basic factors that change across locales. Government sponsorships and support projects can likewise affect the reception and maintainability of lift water system rehearses, for certain locales seeing hearty government drives to advance proficient water use in agribusiness.

Challenges in lift water system rehearses are general, yet their temperament and force vary across districts. In certain areas, the absence of solid admittance to energy sources represents a critical obstacle, restricting the reception of lift water system advances. Upkeep and functional difficulties, including the accessibility of extra parts and specialized mastery, can likewise influence the supportability of lift water system frameworks. Also, the erratic idea of environmental change presents an extra layer of intricacy, with modified precipitation examples and outrageous climate occasions influencing the accessibility and conveyance of water assets.

The job of advancement in lift water system rehearses is apparent in the different arrangements that have arisen to address difficulties and improve proficiency. Mechanical headways, for example, the advancement of more energy-proficient siphons, sensor-based water system frameworks, and brilliant water the executives instruments, add to the development of lift water system rehearses. Advancements in materials and development procedures have likewise prompted the production of more strong and practical lift water system framework.

A similar investigation of lift water system rehearses reaches out past the innovative and functional viewpoints to incorporate the more extensive financial ramifications. The effect on agrarian efficiency, occupations of cultivating networks, and the generally speaking financial improvement of districts is formed by the adequacy and manageability of lift water system frameworks. In locales where effective lift water system rehearses are coordinated into extensive water the board techniques, the advantages are reflected in expanded crop yields, further developed food security, and upgraded financial open doors for ranchers.

In any case, the advantages of lift water system are not uniform across all districts, and abberations in admittance to water assets can compound existing financial disparities. The requirement for comprehensive and impartial water circulation turns into an essential thought in the near examination, provoking conversations on administration structures, strategy systems, and local area support in water the board dynamic cycles.

The relative examination of lift water system rehearses additionally dives into the versatility of these frameworks notwithstanding advancing difficulties, including populace development, urbanization, and changing area use designs. The capacity of lift water system practices to adjust to these unique circumstances is a critical determinant of their drawn out suitability and commitment to reasonable horticulture. Coordinating lift water system into more extensive land-use arranging and water asset the board becomes basic for guaranteeing its proceeded with significance in the consistently changing scene of agribusiness and water accessibility.

All in all, a near examination of lift water system rehearses in various locales uncovers a complicated transaction of topographical, mechanical, ecological, social, financial, and social variables. The variety in approaches mirrors the versatile idea of lift water system to nearby circumstances, underlining the requirement for setting explicit arrangements that offset mechanical advancement with natural maintainability and financial value. As the world countenances mounting difficulties connected with water shortage, environmental change, and food security, the illustrations gathered from a relative investigation of lift water system practices can illuminate all encompassing and strong ways to deal with horticultural water the executives on a worldwide scale.

7.2 Exploration of international collaborations and knowledge exchange

The scene of information and development has gone through a significant change in the time of globalization. The interconnectedness of the world has led to an organization of coordinated efforts and information trade that rises above geological limits. This investigation digs into the elements, advantages, difficulties, and future possibilities of global joint efforts and information trade, looking at how these collaborations shape the headway of science, innovation, culture, and society.

Global joint efforts and information trade are fundamental to the advancement of different fields, going from logical exploration and innovation improvement to social trade and instructive drives. The globalization of information has empowered scientists, scholastics, and experts to interface, team up, and share thoughts on a worldwide scale, encouraging a rich embroidery of culturally diverse impacts that push development and inventiveness.

Logical examination stands apart as a noticeable field where worldwide joint efforts have turned into the standard as opposed to the special case. The intricacy of contemporary logical difficulties frequently requires aptitude from different disciplines and areas.

Cooperative exploration tries unite researchers and scientists with integral abilities and points of view, prompting a more complete comprehension of mind boggling peculiarities.

One striking illustration of global cooperation in logical exploration is the Enormous Hadron Collider (LHC) at CERN (European Association for Atomic Exploration). The LHC, arranged on the line among Switzerland and France, includes large number of researchers from around the world cooperating to investigate major inquiries concerning the idea of the universe. This cooperative exertion represents the aggregate quest for information on a worldwide scale, rising above public limits in the journey for logical disclosure.

Notwithstanding central examination, worldwide coordinated efforts likewise assume a critical part in tending to worldwide difficulties, for example, environmental change, general wellbeing emergencies, and feasible turn of events. Drives like the Intergovernmental Board on Environmental Change (IPCC) unite researchers, policymakers, and specialists from different nations to survey and incorporate logical information, adding to the worldwide discourse on environment activity and strategy detailing.

The advantages of global coordinated efforts stretch out past logical examination to innovation advancement and development. The interconnected idea of the worldwide economy has filled cooperative endeavors in propelling innovation and driving development. Worldwide enterprises regularly participate in cross-line coordinated efforts, shaping innovative work associations to use different skill and assets.

Open-source programming improvement is another domain where worldwide coordinated efforts flourish. Projects like the Linux working framework and the Apache web server include commitments from a worldwide local area of engineers. This cooperative methodology speeds up the speed of programming improvement as well as cultivates a culture of information sharing and straightforwardness that supports the open-source development.

The social and instructive elements of worldwide joint efforts add to the advancement of social orders and the cultivating of worldwide citizenship. Scholastic organizations all over the planet take part in understudy and staff trade programs, cooperative examination drives, and joint degree programs, establishing a powerful climate where thoughts, viewpoints, and information stream uninhibitedly across borders.

Social trade programs, worked with by global coordinated efforts, advance common comprehension and appreciation among assorted networks. Drives like creative trades, writing celebrations, and film coordinated efforts exhibit the extravagance and variety of worldwide societies, separating generalizations and encouraging a feeling of shared mankind.

Regardless of the horde benefits, worldwide joint efforts and information trade accompany their arrangement of difficulties. One critical test is the imbalance in assets and aptitude among teaming up accomplices. Created nations frequently have more noteworthy examination framework, financing, and mechanical capacities, making a power dynamic that can upset evenhanded coordinated efforts. Endeavors to address this lopsidedness require deliberate drives to construct limit in creating locales and guarantee fair and comprehensive support in cooperative undertakings.

Social contrasts and correspondence hindrances can likewise present difficulties to worldwide joint efforts. Dissimilar social standards, correspondence styles, and language boundaries might obstruct compelling information trade. Fruitful coordinated efforts require the improvement of social capability, compelling correspondence procedures, and a guarantee to understanding and regarding different viewpoints.

Licensed innovation privileges and legitimate systems present one more layer of intricacy in worldwide coordinated efforts. Issues connected with possession, licensing, and commercialization of exploration results can be wellsprings of pressure among partners. Laying out clear arrangements and structures for protected innovation the board is fundamental to explore these intricacies and guarantee that the advantages of cooperative endeavors are fairly shared.

International contemplations and global relations can impact the elements of information trade and coordinated efforts. Political strains and discretionary issues between nations might influence the progression of data, the portability of scientists, and the general attainability of cooperative drives. Exploring these international difficulties requires a fragile harmony between logical pursuits and discretionary contemplations.

The continuous Coronavirus pandemic has highlighted both the significance and weakness of worldwide joint efforts even with worldwide emergencies. While the pandemic provoked extraordinary degrees of logical coordinated effort to grasp the infection, foster antibodies, and offer basic data, it additionally uncovered weaknesses in the worldwide wellbeing foundation and featured the requirement for reinforced global participation in tending to wellbeing crises.

Looking forward, the fate of worldwide coordinated efforts and information trade holds both commitment and intricacy. Propels in innovation, especially in correspondence and cooperation apparatuses, are probably going to additionally work with far off joint efforts and extension geological distances. Virtual meetings, online exploration stages, and cooperative programming apparatuses offer new roads for information trade, empowering scientists and experts to cooperate consistently across borders.

The job of non-legislative associations (NGOs), charitable establishments, and global foundations is vital in encouraging and supporting worldwide joint efforts. These elements frequently assume a key part in financing cooperative undertakings, giving foundation support, and working with organizations that address worldwide difficulties. Fortifying the job of such associations can add to a more hearty and comprehensive worldwide information environment.

Training stays a foundation of worldwide coordinated efforts, with organizations and policymakers perceiving the significance of getting ready understudies for a globalized world. Internationalization of training includes integrating worldwide points of view into educational programs, advancing understudy trades, and working with cooperative exploration open doors. This approach furnishes the cutting edge with the abilities and mentality expected to explore and add to a world described by interconnected information frameworks.

The moral components of worldwide joint efforts merit cautious thought. As scientists and establishments participate in cooperative activities, moral guidelines should be maintained to guarantee the respectability of examination results and the security of members. Issues, for example, research morals, information sharing, and social responsiveness require progressing consideration and discourse to encourage a worldwide exploration local area grounded in moral standards.

All in all, the investigation of worldwide coordinated efforts and information trade uncovers a unique scene described by development, challenges, and the common quest for understanding and progress. Whether in the domains of logical exploration, innovation advancement, social trade, or training, cooperative endeavors have become irreplaceable in tending to worldwide difficulties and propelling the boondocks of information. As we explore the intricacies of a globalized world, cultivating comprehensive, fair, and moral global coordinated efforts is fundamental for tackling the aggregate insight of mankind and molding a future set apart by shared thriving and common comprehension

7.3 Showcase of innovative approaches to lift irrigation on a global scale

The domain of lift water system, a basic feature of rural water the board, has seen a range of inventive methodologies across the globe. These creative arrangements plan to improve productivity, maintainability, and availability in giving water to crops. This grandstand investigates different methodologies utilized in lift water system frameworks, traversing mechanical advancements, natural contemplations, local area commitment, and the coordination of environmentally friendly power sources.

Mechanical development stands apart as a foundation of modernizing lift water system frameworks. Propels in siphon innovation, sensor frameworks, and computerization have by and large altered the productivity and accuracy of water conveyance to farming fields.

Variable recurrence drives (VFDs) coordinated into electric siphons consider better control of water stream, coordinating it with the particular requirements of harvests and limiting wastage.

In certain districts, the reception of accuracy water system advancements goes past VFDs. Sensor-based frameworks, utilizing information on soil dampness levels, weather patterns, and yield water prerequisites, empower ranchers to arrive at informed conclusions about when and how much water to apply. This information driven approach improves water use, upgrades crop yields, and adds to asset preservation.

Moreover, the combination of brilliant water system frameworks has gotten some decent forward movement in lift water system rehearses universally. These frameworks frequently influence the Web of Things (IoT) to empower continuous checking and control of water system processes. Ranchers can remotely deal with their lift water system frameworks, getting information and cautions on their cell phones or PCs. This network upgrades the responsiveness of the framework to changing circumstances and further develops generally speaking water use proficiency.

Notwithstanding mechanical developments, the ecological supportability of lift water system frameworks is a vital concentration in numerous worldwide drives. Customary lift water system frameworks might act difficulties such like energy failure, ecological effect, and over-extraction of water assets. Manageable methodologies look to address these worries by integrating eco-accommodating advancements and preservation rehearses.

One eminent development is the reconciliation of sunlight based fueled lift water system frameworks. Sun powered energy gives a sustainable and harmless to the ecosystem option in contrast to customary energy sources. In locales with bountiful daylight, sun oriented siphons have arisen as a practical and manageable arrangement. These frameworks diminish the carbon impression as well as proposition monetary advantages by bringing down functional expenses and making lift water system more open to off-matrix and distant regions.

Additionally, the idea of energy-effective lift water system lines up with the more extensive development towards green advances. Endeavors to improve the energy proficiency of siphons, investigate elective energy sources, and take on energy-saving practices add to the general maintainability of lift water system frameworks. At times, the utilization of energy-effective siphons, like those satisfying the guidelines of the Worldwide Electrotechnical Commission (IEC), has turned into a benchmark for feasible lift water system rehearses.

Natural supportability likewise includes water preservation measures. Effective channel lining, taking on dribble or sprinkler water system techniques, and executing water reaping procedures are among the methodologies utilized to limit water wastage. The accentuation on practical water use lines up with the worldwide basic to address water shortage and guarantee the drawn out reasonability of horticulture.

Local area commitment and participatory methodologies are basic parts of imaginative lift water system rehearses. Perceiving the significance of nearby information and including networks in dynamic cycles add to the achievement and supportability of lift water system drives. Local area oversaw lift water system frameworks engage ranchers to take responsibility for assets, cultivating a feeling of obligation and participation.

Now and again, ranchers' cooperatives or water client affiliations effectively partake in the preparation, execution, and the board of lift water system projects. These cooperative models not just guarantee the impartial dispersion of water assets yet additionally fabricate a feeling of shared liability regarding the upkeep and manageability of the framework.

The progress of local area drove approaches is exemplified in projects where ranchers contribute both monetarily and through work, showing an aggregate obligation to the outcome of the lift water system framework. This model not just improves the financial reasonability of the drive yet additionally reinforces social attachment among local area individuals.

Conventional and native information frameworks likewise assume a vital part in forming imaginative ways to deal with lift water system. Drawing on nearby skill, ranchers might execute age-old methods that are appropriate to the particular biological and social setting of their district. These customary techniques, frequently went down through ages, add to the strength and versatility of lift water system rehearses.

Notwithstanding people group commitment, strategy systems and institutional help are vital in advancing creative lift water system draws near. Legislatures and global associations frequently assume a reactant part by giving monetary motivating forces, specialized skill, and administrative structures that energize the reception of manageable and proficient lift water system frameworks.

For example, endowments for the establishment of sunlight based fueled siphons, monetary help for water protection drives, and strategies advancing the utilization of energy-productive advancements all add to the far reaching reception of imaginative lift water system rehearses. Clear rules on water designation, supportable practices, and participatory administration structures further improve the effect of strategy mediations.

On a worldwide scale, cooperative drives and information trade stages work with the cross-fertilization of thoughts and encounters connected with lift water system. Global associations, research foundations, and NGOs add to a common store of best practices, contextual investigations, and specialized skill that benefit locales all over the planet confronting comparable difficulties.

The reception of nature-based arrangements is acquiring conspicuousness as an imaginative way to deal with lift water system. Environment based approaches influence the normal elements of biological systems to upgrade water accessibility and quality. Practices like afforestation, wetland rebuilding, and soil protection add to further developed water maintenance and diminished sedimentation, emphatically affecting lift water system frameworks.

Moreover, coordinating lift water system with agroecological standards lines up with the more extensive development towards practical and regenerative agribusiness. Agroecology underlines the amicable concurrence of agribusiness and the climate, advancing practices that upgrade biodiversity, soil wellbeing, and water preservation. Lift water system frameworks planned with agroecological standards focus on natural equilibrium and strength.

In dry and semi-dry locales, advancements in water reaping are changing the water scene. Water reaping methods, for example, check dams, roof water collecting, and shape digging, catch and store water, enhancing customary water hotspots for lift water system. These decentralized and nature-based arrangements add to water security and strength in water-pushed locales.

The combination of information examination and computerized reasoning (artificial intelligence) is introducing another period of accuracy farming inside the area of lift water system. Man-made intelligence driven models investigate enormous datasets, including environment designs, soil conditions, and harvest execution, to enhance water system timetables and asset assignment. This information driven approach upgrades the accuracy and effectiveness of lift water system, guaranteeing that water assets are applied prudently to expand farming efficiency.

In districts confronting water shortage, creative desalination advances are being investigated to change over seawater into a suitable hotspot for lift water system. Desalination, frequently fueled by environmentally friendly power sources, offers a likely answer for water shortage challenges in seaside regions. While the innovation is as yet advancing, effective executions exhibit the extraordinary capability of bridling unusual water hotspots for farming use.

Arising advances like blockchain are additionally tracking down applications in the domain of lift water system. Blockchain innovation works with straightforward and secure exchanges, which can be utilized for effective water the executives. Savvy contracts on blockchain stages can mechanize cycles like water distribution, charging, and support, smoothing out the authoritative parts of lift water system frameworks.

The job of ladies in farming water the executives, including lift water system, is earning respect as a significant part of advancement. Ladies frequently assume focal parts in water assortment, the board, and use in agrarian social orders. Engaging ladies with information, abilities, and admittance to assets adds to more comprehensive and maintainable lift water system rehearses.

Creative funding models are arising to address the capital-serious nature of lift water system projects. Public-private associations, influence speculations, and crowd-funding drives offer elective roads for subsidizing, diminishing the monetary weight on individual ranchers or legislatures. These creative funding instruments upgrade the possibility and versatility of lift water system projects, guaranteeing that they contact a more extensive crowd.

Lift water system, a fundamental part of horticultural water the board, assumes a crucial part in guaranteeing water access for crops in districts where normal water sources are not effectively open or adequate. The worldwide scene of lift water system is different and dynamic, molded by topographical varieties, mechanical advancements, natural contemplations, and financial elements. This investigation dives into the subtleties of lift water system on a worldwide scale, looking at its importance, challenges, creative arrangements, and the more extensive ramifications for maintainable water use in farming.

Meaning of Lift Water system:

Lift water system holds huge importance in locales where the normal angle of land isn't helpful for gravity-based water system frameworks. In bumpy or hilly landscapes, where water sources are frequently situated at lower heights, lift water system turns into a urgent arrangement. The interaction includes lifting water from lower levels to higher heights, guaranteeing that farming terrains arranged uphill get the imperative water system.

The significance of lift water system is complemented by its job in alleviating water shortage challenges. In parched and semi-bone-dry districts, where water assets are restricted, lift water system gives a life saver to farming. By taking advantage of elective water sources and utilizing productive lifting systems, ranchers can develop crops even in regions customarily viewed as unacceptable for farming because of water limitations.

The versatility of lift water system frameworks is apparent in their application across different scenes, from the fields to hilly locales. In level landscapes, lift water system supplements gravity-based frameworks by tending to explicit difficulties like lopsided water dispersion, profound water tables, or the need to get to water from modern sources. This flexibility highlights the strength and utility of lift water system in gathering the differed needs of farming around the world.

Worldwide Varieties in Lift Water system Practices:

The elements of lift water system rehearses differ altogether across districts, mirroring the many-sided exchange of geological, mechanical, and financial variables. In mechanically progressed economies, electric siphons are regularly utilized for lifting water, utilizing solid admittance to power. Conversely, in off-matrix or distant regions, diesel-fueled siphons or elective energy sources, for example, sun powered or wind energy arise as viable arrangements.

Also, the decision of lifting system is in many cases affected by financial contemplations. In districts with plentiful and reasonable power, electric siphons might be the favored choice. On the other hand, in regions confronting energy difficulties or high fuel costs, ranchers might choose more financially savvy and manageable other options, for example, sun based controlled siphons, which have acquired noticeable quality as a cleaner and monetarily suitable choice.

The wellsprings of water for lift water system likewise display varieties internationally. A few locales depend on surface water bodies like streams, lakes, or repositories, while others tap into groundwater through wells or boreholes. The supportability of lift water system rehearses is complicatedly connected to the sensible administration of these water sources, stressing the requirement for far reaching water asset arranging and protection measures.

The natural effect of lift water system rehearses is a critical thought in the worldwide setting. Extricating water from regular sources can have biological outcomes, remembering changes for stream environments, soil debasement, and possible struggles over water use. Reasonable lift water system rehearses consolidate measures to limit ecological effect, for example, the utilization of water-saving innovations, re-energizing springs, and embracing soil preservation rehearses.

Mechanical Developments in Lift Water system:

Mechanical development stands apart as a main thrust in the development of lift water system frameworks around the world. The approach of cutting edge siphoning advancements, robotization, and brilliant water system arrangements has changed the productivity and accuracy of water conveyance to agrarian fields.

Variable Recurrence Drives (VFDs) incorporated into electric siphons address a critical innovative progression. VFDs consider better control of water stream, empowering ranchers to coordinate water system rates with explicit yield needs. This degree of accuracy advances water use as well as adds to energy productivity by changing siphon speed in view of interest.

Sensor-based water system frameworks address one more jump forward in lift water system innovation. These frameworks use information from sensors estimating soil dampness levels, weather patterns, and harvest water necessities. The information driven approach works with informed direction, permitting ranchers to fit water system timetables to the particular necessities of their harvests. Such accuracy in water application improves farming efficiency while saving water assets.

Shrewd water system frameworks, frequently utilizing the Web of Things (IoT), have acquired unmistakable quality in lift water system rehearses.

These frameworks empower continuous observing and control of water system processes, engaging ranchers to remotely deal with their frameworks. The network managed the cost of by shrewd frameworks improves responsiveness to evolving conditions, adding to more proficient and supportable water use in farming.

Besides, the coordination of Computerized reasoning (artificial intelligence) and information examination is introducing another time of accuracy agribusiness inside lift water system. Simulated intelligence driven models break down tremendous datasets, including environment designs, soil conditions, and yield execution, to improve water system timetables and asset distribution. This information driven approach improves the accuracy and proficiency of lift water system, guaranteeing that water assets are applied wisely to amplify agrarian efficiency.

Natural Supportability and Sustainable power:

The basic of natural supportability has provoked creative arrangements in lift water system, especially in the domain of fuel sources. The conventional dependence on petroleum derivatives for fueling siphons has given way to a flood of environmentally friendly power arrangements, adjusting lift water system practices to more extensive manageability objectives.

Sun oriented controlled lift water system frameworks have arisen as an extraordinary arrangement, particularly in locales with plentiful daylight. Sun based siphons offer a reasonable and savvy option in contrast to traditional energy sources. These frameworks lessen ozone harming substance emanations as well as add to the decentralization of force age, making lift water system more available to off-lattice and distant regions.

The ecological manageability of lift water system is additionally highlighted by water protection measures. Proficient channel lining, reception of dribble or sprinkler water system strategies, and water collecting procedures all add to limiting water wastage. Manageable practices that focus on biological equilibrium, soil wellbeing, and biodiversity preservation line up with worldwide endeavors to address water shortage and environmental change.

Local area Commitment and Participatory Methodologies:

The achievement and manageability of lift water system drives rely on local area commitment and participatory methodologies. Perceiving the significance of neighborhood information and including networks in dynamic cycles are fundamental components of compelling lift water system rehearses.

In various occurrences, ranchers' cooperatives or water client affiliations effectively partake in the preparation, execution, and the board of lift water system projects. These cooperative models guarantee the impartial conveyance of water assets as well as construct a feeling of shared liability regarding the upkeep and manageability of the foundation.

The outcome of local area drove approaches is clear in projects where ranchers contribute both monetarily and through work, cultivating an aggregate obligation to the progress of the lift water system framework.

Customary and native information frameworks likewise assume a vital part in forming imaginative ways to deal with lift water system. Drawing on neighborhood mastery, ranchers might execute age-old strategies that are appropriate to the particular biological and social setting of their locale. These customary strategies, frequently went down through ages, add to the strength and flexibility of lift water system rehearses.

Strategy Systems and Institutional Help:

Strategy systems and institutional help assume an essential part in advancing imaginative lift water system draws near. States, global associations, and research organizations frequently give monetary motivating forces, specialized skill, and administrative structures that support the reception of economical and effective lift water system frameworks.

For example, endowments for the establishment of sunlight based controlled siphons, monetary help for water preservation drives, and strategies advancing the utilization of energy-productive advancements all add to the boundless reception of inventive lift water system rehearses. Clear rules on water portion, feasible practices, and participatory administration structures further improve the effect of strategy mediations.

Difficulties and Future Possibilities:

Notwithstanding the advancement in lift water system rehearses universally, challenges persevere, mirroring the intricacy of water the executives in agribusiness. One huge test is the imbalance in assets and skill among working together accomplices. Created nations frequently have more prominent examination foundation, subsidizing, and mechanical capacities, making a power dynamic that can prevent fair joint efforts. Endeavors to address this irregularity require purposeful drives to fabricate limit in creating districts and guarantee fair and comprehensive cooperation in cooperative undertakings.

Social contrasts and correspondence boundaries can likewise present difficulties to worldwide coordinated efforts in lift water system. Dissimilar social standards, correspondence styles, and language boundaries might obstruct compelling information trade. Fruitful joint efforts require the improvement of social capability, viable correspondence techniques, and a pledge to understanding and regarding different viewpoints.

Licensed innovation freedoms and lawful systems present one more layer of intricacy in lift water system joint efforts. Issues connected with proprietorship, protecting, and commercialization of exploration results can be wellsprings of strain among teammates.

Laying out clear arrangements and systems for protected innovation the executives is fundamental to explore these intricacies and guarantee that the advantages of cooperative endeavors are evenhandedly shared.

International contemplations and global relations can impact the elements of information trade and coordinated efforts in lift water system. Political strains and conciliatory issues between nations might influence the progression of data, the portability of analysts, and the general possibility of cooperative drives. Exploring these international difficulties requires a fragile harmony between logical pursuits and political contemplations.

The continuous Coronavirus pandemic has highlighted both the significance and weakness of worldwide joint efforts even with worldwide emergencies. While the pandemic provoked remarkable degrees of logical coordinated effort to figure out the infection, foster immunizations, and offer basic data, it additionally uncovered weaknesses in the worldwide wellbeing foundation and featured the requirement for reinforced global collaboration in tending to wellbeing crises.

Looking forward, the eventual fate of global coordinated efforts and information trade in lift water system holds both commitment and intricacy. Progresses in innovation, especially in correspondence and coordinated effort apparatuses, are probably going to additionally work with far off joint efforts and extension geological distances. Virtual meetings, online examination stages, and cooperative programming instruments offer new roads for information trade, empowering specialists and experts to cooperate consistently across borders.

The job of non-legislative associations (NGOs), magnanimous establishments, and global foundations is significant in encouraging and supporting worldwide coordinated efforts in lift water system. These substances frequently assume a key part in financing cooperative ventures, giving framework support, and working with organizations that address worldwide difficulties. Reinforcing the job of such associations can add to a more vigorous and comprehensive worldwide information environment.

Schooling stays a foundation of global joint efforts in lift water system, with establishments and policymakers perceiving the significance of getting ready understudies for a globalized world. Internationalization of instruction includes integrating worldwide viewpoints into educational plans, advancing understudy trades, and working with cooperative exploration valuable open doors. This approach outfits the cutting edge with the abilities and mentality expected to explore and add to a world described by interconnected information frameworks.

The moral elements of worldwide joint efforts in lift water system merit cautious thought. As specialists and foundations participate in cooperative activities, moral norms should be maintained to guarantee the respectability of exploration results and the security of members. Issues, for example, research morals, information sharing, and social responsiveness require progressing consideration and exchange to encourage a worldwide examination local area grounded in moral standards.

Chapter 8

Future Trends and Innovations

What's to come is a consistently advancing material, painted by the brushstrokes of advancement and driven by the unavoidable trends. As we explore the unfamiliar regions of tomorrow, a few patterns and developments arise, molding the scene of ventures, social orders, and human life. From the domains of innovation to the subtleties of medical services, what's to come guarantees an embroidery woven with headways that reclassify how we might interpret what is conceivable.

One of the conspicuous strings in this unpredictable embroidered artwork is the proceeded with development of man-made consciousness (computer based intelligence). Simulated intelligence, when an idea consigned to the domain of sci-fi, has turned into a basic piece of our regular routines. The fate of artificial intelligence holds the commitment of considerably more noteworthy steps, as scientists and designers push the limits of AI and mental figuring. Profound learning calculations, motivated by the brain organizations of the human mind, are at the front of this man-made intelligence unrest.

Before very long, we can anticipate that artificial intelligence should pervade different features of our reality, affecting all that from customized medical care to independent vehicles. The capacity of man-made intelligence frameworks to examine immense datasets and extricate significant experiences will assume an essential part in clinical diagnostics. Prescient investigation, controlled by artificial intelligence, will empower medical care experts to recognize potential wellbeing chances and intercede proactively. This shift towards precaution and customized medication can possibly reform the medical care industry, making therapies more powerful and diminishing the weight on medical care frameworks.

Independent vehicles, another field where artificial intelligence is utilizing its muscles, are ready to reshape the manner in which we drive. The assembly of simulated intelligence, sensors, and availability is moving the advancement of self-driving vehicles. These vehicles, furnished with refined simulated intelligence calculations, can explore complex traffic situations, pursue split-subsequent options, and speak with one another to improve traffic stream. The broad reception of independent vehicles holds the commitment of diminishing mishaps, further developing traffic productivity, and changing transportation into a more secure and more consistent experience.

The computerized upset isn't restricted to simulated intelligence alone; blockchain innovation is arising as a groundbreaking power with sweeping ramifications. At first known as the basic innovation for digital currencies like Bitcoin, blockchain has developed into a decentralized record framework with applications across different enterprises. The permanence and straightforwardness of blockchain make it an optimal contender for guaranteeing the respectability of information in fields, for example, finance, production network, and medical care.

In finance, blockchain can possibly change how exchanges are directed. Cryptographic forms of money, based on blockchain innovation, are testing conventional thoughts of cash and money. The decentralized idea of blockchain dispenses with the requirement for delegates, decreasing exchange costs and speeding up monetary exchanges. National banks and monetary organizations are investigating the chance of giving their own advanced monetary forms, utilizing the proficiency and security presented by blockchain.

The production network is another field where blockchain is making critical advances. The detectability and straightforwardness gave by blockchain can resolve issues, for example, fake merchandise and store network misrepresentation. By recording each exchange on a changeless record, partners can follow the excursion of items from producer to purchaser, guaranteeing legitimacy and quality. This improves buyer trust as well as smoothes out store network tasks.

Medical care, with its accentuation on information respectability and patient security, stands to profit from the execution of blockchain innovation. Electronic wellbeing records (EHRs) put away on a blockchain can give a protected and carefully designed vault of patient data. This not just guarantees the privacy of touchy clinical information yet in addition works with interoperability among various medical services suppliers, working on the coherence of care.

The appearance of 5G innovation is one more impetus for extraordinary change not too far off. The fifth era of remote correspondence vows to alter network, empowering quicker web speeds, lower inertness, and expanded limit with regards to gadgets. The ramifications of 5G stretch out past quicker download speeds on cell phones; they envelop a change in perspective in the manner we collaborate with innovation.

The Web of Things (IoT) is one of the essential recipients of 5G network. The capacity of 5G organizations to deal with an enormous number of associated gadgets all the while makes the way for another time of shrewd urban communities, savvy homes, and modern mechanization. The consistent correspondence between IoT gadgets worked with by 5G will make an organized biological system where gadgets can team up continuously, making our surroundings more proficient, responsive, and feasible.

The mix of 5G and IoT is especially extraordinary with regards to shrewd urban communities. These metropolitan conditions influence associated sensors and gadgets to accumulate continuous information on different perspectives, for example, traffic stream, energy utilization, and air quality. The information gathered is then dissected to advance city tasks, upgrade public administrations, and work on the general personal satisfaction for inhabitants. From astute traffic the executives to garbage removal, savvy urban communities embody the capability of 5G and IoT to make more bearable and manageable metropolitan spaces.

The convergence of innovation and manageability is a point of convergence in conversations about what's in store. As worries about environmental change increase, advancements that advance natural maintainability become the overwhelming focus. Environmentally friendly power sources, for example, sun based and wind power, are encountering quick progressions, making them more feasible options in contrast to customary petroleum products.

The broad reception of electric vehicles (EVs) is a substantial sign of the shift towards supportability in transportation. As battery innovation improves and charging framework turns out to be more inescapable, EVs are turning into a standard decision for earth cognizant customers. Legislatures and businesses are putting vigorously in the advancement of EVs to decrease fossil fuel byproducts and relieve the effect of environmental change.

Past transportation, supportable practices are getting momentum in different businesses. The idea of a round economy, where assets are reused, reused, and reused, is picking up speed. Developments in materials science and reusing advances are driving the change towards a more economical and round way to deal with creation and utilization.

The combination of biotechnology and data innovation is leading to another period of customized medication. Progresses in genomics, combined with the force of enormous information examination, empower medical services suppliers to fit therapies to the singular qualities of patients. Accuracy medication, as it is frequently called, addresses a takeoff from the one-size-fits-all way to deal with medical services, introducing a period where treatments are modified in light of an individual's hereditary cosmetics, way of life, and ecological variables.

The capacity to grouping and examine the human genome at an uncommon scale is a foundation of the customized medication transformation. The expense of genomic sequencing has dove throughout the long term, making it more available for both exploration and clinical applications. This has prepared for forward leaps in grasping the hereditary premise of sicknesses and distinguishing designated treatments.

CRISPR-Cas9, a progressive quality altering innovation, adds one more layer to the customized medication scene. CRISPR permits researchers to change the DNA of living creatures, making the way for the amendment of hereditary imperfections and the advancement of novel restorative intercessions unequivocally. While the moral ramifications of quality altering are a subject of progressing banter, the possibility to destroy hereditary illnesses and upgrade human capacities is a tempting possibility.

The fate of work is going through a change in outlook, driven by the convergence of computerization, man-made intelligence, and changing cultural perspectives. The conventional idea of an all day office work is developing into an additional adaptable and decentralized model. Remote work, when a need forced by worldwide occasions, is turning into a long-lasting component of the cutting edge work scene.

The ascent of the gig economy and the pervasiveness of independent work are reshaping the business scene. Stages that interface consultants with clients are multiplying, furnishing people with the adaptability to pick when and where they work. While this offers newly discovered independence for laborers, it likewise brings up issues about professional stability, benefits, and the general design of the work market.

Computerization, fueled by artificial intelligence and mechanical technology, is robotizing routine assignments across different businesses. While this computerization improves effectiveness and efficiency, it additionally raises worries about work removal.

The developing idea of work requires an emphasis on reskilling and upskilling the labor force to flourish in a computerized economy. Long lasting acquiring and versatility become critical abilities in a time where the half-existence of abilities is diminishing, and the interest for new capacities is on the ascent.

The idea of expanded reality (AR) and augmented reality (VR) is rising above the domain of gaming and amusement, tracking down applications in assorted fields. AR overlays computerized data onto the actual world, improving our insight and cooperation with the climate. VR, then again, submerges clients in a PC produced climate, making a mimicked reality.

In medical care, AR and VR are changing clinical preparation and patient consideration. Specialists can utilize AR to picture basic data, like the area of veins, during methodology. VR is utilized in openness treatment for psychological well-being conditions, permitting patients to stand up to and defeat fears in a controlled virtual climate. The utilization of these advancements stretches out to schooling, where AR and VR upgrade opportunities for growth by giving vivid recreations and intuitive substance.

In the domain of amusement, the limits between the virtual and actual universes are obscuring. Broadened reality (XR), a term that envelops AR, VR, and blended the truth, is molding the future of narrating and gaming. XR encounters offer clients a degree of drenching and intuitiveness that was once the stuff of sci-fi. As innovation progresses, the line between the genuine and the virtual will keep on obscuring, opening up additional opportunities for imagination and articulation.

The journey for manageability stretches out past the actual world to the computerized domain. The ecological effect of the quickly extending computerized framework, from server farms to electronic gadgets, is a developing concern. The energy utilization related with the activity of these computerized frameworks adds to fossil fuel byproducts and natural debasement.

Endeavors are in progress to foster more energy-productive advancements and supportable practices in the computerized space. Green figuring drives center around decreasing the carbon impression of server farms through developments in cooling frameworks, energy-effective equipment, and the utilization of environmentally friendly power sources. The idea of round registering, roused by the standards of the roundabout economy, advocates for broadening the life expectancy of electronic gadgets through fix, reuse, and reusing.

The moral elements of innovation have come to the very front as advancements outperform the improvement of administrative systems and moral rules. Issues like security, predisposition in artificial intelligence calculations, and the moral ramifications of quality altering present complex difficulties that require smart thought. The mindful turn of events and sending of innovation request a cooperative exertion including scientists, policymakers, industry pioneers, and society at large.

Security concerns pose a potential threat in a time of omnipresent network and information assortment. The tremendous measure of individual information created by advanced communications presents dangers to individual security on the off chance that not dealt with dependably. Finding some kind of harmony between the advantages of information driven developments and the security of protection is quite difficult for policymakers and innovation organizations.

Predisposition in artificial intelligence calculations, frequently an impression of the predispositions present in the information used to prepare these frameworks, is a squeezing moral concern. From employing calculations that propagate orientation and racial inclinations to prescient policing frameworks that support existing imbalances, the ramifications of one-sided computer based intelligence are expansive. Addressing predisposition in computer based intelligence requires a coordinated work to work on the variety of datasets, execute straightforward calculations, and lay out moral rules for simulated intelligence improvement.

The moral contemplations encompassing quality altering advancements like CRISPR-Cas9 bring up basic issues about the limits of human mediation in the regular request. The possibility to alter the human germline, consequently affecting the hereditary cosmetics of people in the future, raises significant moral, social, and philosophical inquiries. As these innovations advance, moral structures should develop to guarantee their dependable and evenhanded use.

What's in store unfurls at the crossing point of mechanical advancement and human creativity. Exploring this territory requires an all encompassing viewpoint that thinks about the innovative headways themselves as well as their cultural, moral, and ecological ramifications. As we stand on the slope of another time, the decisions today will shape the direction of tomorrow. What's in store is certainly not a foreordained objective; it is an aggregate creation, formed by the activities and choices of people, networks, and social orders all over the planet. Embracing the potential open doors and difficulties that lie ahead, we set out on an excursion of investigation, revelation, and change, graphing a course towards a future that mirrors the best of our yearnings and the versatility of the human soul.

8.1 Anticipation of emerging technologies in lift irrigation

The expectation of arising advancements in lift water system proclaims another time in maintainable water the executives and farming practices. Lift water system, a technique for raising water from a lower to a higher height, is a basic part of horticulture, particularly in locales where water shortage represents a critical test. As the world wrestles with the effects of environmental change and expanding requests on water assets, the joining of inventive innovations into lift water system frameworks holds the commitment of upgrading proficiency, preserving water, and relieving the ecological impression.

One of the key regions where arising advances are making progress in lift water system is in the domain of accuracy horticulture. Accuracy agribusiness use trend setting innovations like sensors, robots, and information examination to advance the utilization of assets, including water. With regards to lift water system, accuracy horticulture empowers ranchers to unequivocally deal with the use of water, guaranteeing that yields get the perfect sum brilliantly.

Sensors assume a vital part in this accuracy driven approach. Soil dampness sensors, for example, give continuous information on the dampness levels in the dirt. This data permits ranchers to tailor their water system rehearses in view of the genuine necessities of the harvests, staying away from over-water system and limiting water wastage. Incorporating these sensors into lift water system frameworks gives a unique input circle, empowering ranchers to settle on information informed conclusions about when and how much water to lift for water system.

Drones, outfitted with different imaging innovations, offer an elevated perspective of horticultural fields. This airborne point of view permits ranchers to evaluate crop wellbeing, recognize areas of stress or water lack, and screen the general state of the fields. By coordinating robot innovation with lift water system, ranchers gain significant experiences that illuminate water system methodologies. For instance, on the off chance that a robot recognizes explicit regions with water pressure, the lift water system framework can be coordinated to focus on those zones, improving water circulation and advancing harvest wellbeing.

The approach of the Web of Things (IoT) further upgrades the capacities of lift water system frameworks. IoT-empowered gadgets, like brilliant valves and regulators, work with the computerization and remote observing of water system processes. These gadgets can be coordinated into existing lift water system foundation, considering unified control and ongoing changes. Ranchers can remotely get to and deal with their lift water system frameworks, answering immediately to changing circumstances and advancing water use with more prominent effectiveness.

Computerized reasoning (artificial intelligence) calculations, a subset of arising innovations, carry a layer of insight to lift water system. By investigating immense datasets from sensors, weather conditions conjectures, and verifiable water system designs, simulated intelligence can anticipate ideal water system plans. This prescient ability permits lift water system frameworks to work proactively, changing water conveyance in light of expected needs and ecological circumstances. Simulated intelligence driven bits of knowledge can assist ranchers with streamlining their water use, decrease functional expenses, and improve by and large agrarian efficiency.

Energy effectiveness is a basic thought in lift water system, where the lifting of water to raised landscapes frequently requires significant energy inputs. Arising advancements offer answers for address this test and make lift water system more maintainable.

Sun based controlled lift water system frameworks, for example, tackle energy from the sun to drive siphons and different parts. As the expense of sun based innovation keeps on diminishing, the reception of sun oriented fueled lift water system turns into a financially reasonable and harmless to the ecosystem choice.

Progressions in siphon innovation likewise add to energy proficiency in lift water system. Variable recurrence drives (VFDs) take into consideration the change of siphon speeds in view of genuine interest, lessening energy utilization during times of lower water necessities. Furthermore, advancements in siphon plan, like the improvement of additional effective and sturdy materials, add to the life span and unwavering quality of lift water system frameworks.

Chasing manageability, the joining of water-saving advancements is foremost. Dribble water system, a technique where water is conveyed straightforwardly to the root zone of plants, limits water wastage contrasted with customary flood water system. While not another innovation, the continuous development in dribble water system frameworks improves their effectiveness and versatility to different yields and territories. Incorporating trickle water system with lift frameworks guarantees that the water lifted is utilized with most extreme accuracy, lessening misfortunes because of vanishing and overflow.

Besides, the utilization of dampness responsive water system frameworks supplements the objectives of water preservation. These frameworks, which change water conveyance in view of ongoing soil dampness levels, guarantee that water system is exactly lined up with the necessities of the harvests. When coordinated into lift water system, dampness responsive frameworks improve the general productivity of water dispersion, adding to asset preservation and natural supportability.

Progressions in materials science and designing are additionally impacting the plan and development of lift water system foundation. Lightweight and solid materials, for example, high-strength polymers and composites, offer options in contrast to conventional materials like metal. The utilization of these high level materials not just decreases the weight and cost of parts yet in addition improves their protection from consumption and wear, bringing about longer life expectancies and lower support necessities for lift water system frameworks.

The expectation of arising advancements in lift water system reaches out past the ranch to the more extensive water the executives environment. Savvy water matrices, empowered by sensor organizations and constant information examination, enhance the designation and dissemination of water assets. These insightful frameworks can powerfully change water stream in view of interest, focus on basic requirements, and answer changing natural circumstances. Incorporating lift water system into savvy water lattices makes a more responsive and versatile water the board foundation that benefits both horticultural and metropolitan water clients.

The combination of lift water system with propels in desalination advancements opens up additional opportunities for using modern water sources. Desalination, the most common way of eliminating salt and different debasements from seawater, gives an extra water asset to areas confronting freshwater shortage. By coordinating desalination plants with lift water system frameworks, seaside regions can use seawater assets for horticultural purposes, diminishing reliance on freshwater sources and easing strain on customary water supplies.

The cooperative utilization of satellite innovation and geographic data frameworks (GIS) improves the preparation and the executives of lift water system projects. Satellite symbolism gives nitty gritty data about geography, land use, and vegetation cover, empowering organizers to distinguish appropriate areas for lift framework. GIS innovation helps with spatial examination, considering the ideal plan and situation of lift frameworks in view of elements, for example, landscape, soil types, and harvest circulation. The mix of these innovations smoothes out the preparation and execution of lift water system projects, working on their general viability and limiting ecological effect.

In the domain of strategy and administration, arising advancements offer apparatuses for more successful water the board and assignment. Blockchain, a decentralized and carefully designed record innovation, can be applied to follow water exchanges and freedoms straightforwardly. By making a safe and irrefutable record of water portions, blockchain innovation improves responsibility and decreases the potential for disagreements regarding water freedoms. This is especially important in districts where water shortage and contending requests require clear and enforceable water administration components.

The combination of ongoing checking and information investigation into water administration frameworks further improves their viability. Leaders can access exceptional data on water use, request, and accessibility, taking into consideration more educated and opportune strategy choices. This information driven approach empowers legislatures and water specialists to carry out versatile water the executives procedures that answer changing circumstances and advance the maintainable utilization of water assets.

The expectation of arising advancements in lift water system isn't without its difficulties and contemplations. The reception of these advances requires interest in examination, improvement, and framework. Besides, there is a requirement for limit building drives to guarantee that ranchers and partners are furnished with the information and abilities to use these innovations really. Tending to these difficulties requires a cooperative exertion including legislatures, research foundations, innovation designers, and nearby networks.

Ecological manageability is a center thought in the combination of arising innovations into lift water system. While these advancements offer the possibility to upgrade effectiveness and efficiency, their execution should be directed by rules that focus on natural preservation.

Adjusting the requirement for expanded farming result with the protection of biological systems and water assets is fundamental to guarantee an amicable and manageable future.

Moral contemplations likewise assume a part in the sending of arising advances in lift water system. As information turns into a focal part of accuracy farming and shrewd water the executives, issues of information possession, protection, and security come to the very front.

8.2 Discussion of research and development initiatives in the field

The conversation of innovative work drives in different fields addresses a foundation in the headway of information, innovation, and development. Innovative work (Research and development) assume an essential part in pushing the limits of what is known, investigating new outskirts, and tending to squeezing difficulties across different spaces. From logical forward leaps to mechanical developments, the results of Research and development drives shape the present and characterize the direction representing things to come.

In the domain of medical services, Research and development drives are impetuses for groundbreaking headways that upgrade clinical therapies, analytic apparatuses, and by and large quiet consideration. Biomedical exploration, driven by a mix of scholarly, legislative, and confidential area endeavors, tries to unwind the intricacies of infections, figure out natural cycles, and find novel treatments. One conspicuous model is the continuous exploration in genomics, where researchers endeavor to unravel the human genome and distinguish hereditary variables adding to different illnesses.

Accuracy medication, a result of genomic research, addresses a change in perspective in medical care. Rather than a one-size-fits-all methodology, accuracy medication tailors medicines to the singular qualities of patients, considering their hereditary cosmetics, way of life, and natural variables. Drives, for example, the Accuracy Medication

Drive in the US expect to speed up research in this field, encouraging coordinated effort between scientists, medical services suppliers, and patients to introduce another time of customized and successful clinical mediations.

The field of man-made consciousness (simulated intelligence) is seeing a strong flood of Research and development drives pointed toward propelling AI calculations, regular language handling, and PC vision. These drives length scholarly establishments, tech goliaths, and new businesses, each adding to the advancement of computer based intelligence innovations. Progressing research centers around working on man-made intelligence's capacity to grasp complex examples, make exact expectations, and improve human-PC connections.

Profound learning, a subset of AI roused by brain organizations, has gathered critical consideration as of late. Research and development endeavors in profound learning investigate its applications in assorted regions, from picture and discourse acknowledgment to independent frameworks.

The advancement of additional proficient calculations, combined with the accessibility of huge datasets, has moved profound learning into viable applications, including medical services diagnostics, language interpretation, and independent vehicles.

Quantum processing remains at the front of Research and development drives that guarantee to change computational capacities. Not at all like old style PCs that utilization pieces to address data as either 0s or 1s, quantum PCs influence qubits, which can exist in numerous states at the same time. This intrinsic parallelism empowers quantum PCs to tackle complex issues, for example, improvement and cryptography, a lot quicker than old style PCs.

Worldwide endeavors in quantum processing Research and development include coordinated efforts between state run administrations, scholastic establishments, and industry players. Drives mean to beat the specialized difficulties related with building and keeping up with stable quantum PCs, investigate quantum calculations for commonsense applications, and lay the foundation for a future where quantum figuring turns into a basic piece of logical and computational undertakings.

Manageability and natural preservation are squeezing worldwide worries, and Research and development drives in these spaces center around creating imaginative answers for address environmental change, asset exhaustion, and biological debasement. Green innovations, incorporating environmentally friendly power, energy proficiency, and feasible farming, address a huge area of innovative work.

The mission for spotless and environmentally friendly power sources has prompted leap forwards in sunlight based, wind, and hydropower advancements. Sunlight based photovoltaic cells, for example, keep on going through Research and development to further develop effectiveness, diminish expenses, and upgrade adaptability. Additionally, progressions in energy capacity advancements, like batteries, add to the suitability of environmentally friendly power by tending to irregularity and empowering the capacity of overabundance energy for sometime in the future.

In practical agribusiness, Research and development drives investigate techniques to increment crop yields while limiting ecological effect. Accuracy agribusiness, empowered by advances like IoT, sensors, and information examination, improves asset use, upgrades soil wellbeing, and lessens the environmental impression of cultivating rehearses. Hereditary designing and harvest reproducing are likewise areas of dynamic examination, expecting to foster yields that are more impervious to bugs, infections, and ecological stressors.

Water shortage is a basic worldwide test, and Research and development drives in water the executives and protection look to foster proficient and maintainable arrangements. Developments in water refinement innovations, for example, desalination and high level filtration frameworks, plan to give admittance to clean water in areas confronting deficiencies.

Shrewd water the board frameworks, consolidating IoT sensors and information examination, improve water utilization, decrease wastage, and upgrade generally speaking water asset productivity.

With regards to environmental change relief, Research and development drives investigate carbon catch and capacity advancements, manageable transportation arrangements, and strong framework plan. Research endeavors likewise dig into understanding the effects of environmental change and forming transformation techniques for weak networks and biological systems. Cooperative worldwide drives unite researchers, policymakers, and partners to on the whole address the intricate and interconnected difficulties presented by environmental change.

The field of materials science is a center of Research and development exercises zeroed in on creating progressed materials with novel properties and applications. Nanotechnology, for example, includes controlling materials at the nanoscale to accomplish special qualities, prompting developments in gadgets, medication, and assembling. Graphene, a solitary layer of carbon iotas organized in a hexagonal grid, is a material of serious exploration interest because of its uncommon strength, conductivity, and flexibility.

Metamaterials, designed materials with properties not tracked down in nature, offer open doors for controlling electromagnetic waves, sound, and intensity. Research and development drives in metamaterials investigate applications, for example, shrouding gadgets, superlenses, and high level sensors. The consistent investigation of new materials and the refinement of existing ones add to the advancement of state of the art innovations across different businesses.

Space investigation and stargazing address outskirts where Research and development drives push the limits of human information and innovative abilities. Space organizations, research foundations, and privately owned businesses team up on projects going from planetary investigation to the investigation of far off systems. The journey for grasping the beginnings of the universe, the quest for extraterrestrial life, and the investigation of heavenly bodies drive progressing research drives.

Mars investigation has been a point of convergence of ongoing Research and development endeavors, with mechanical missions intended to concentrate on the Martian surface and quest for indications of past or present life. The improvement of cutting edge rocket, impetus frameworks, and independent innovations adds to the achievability and outcome of interplanetary investigation. Global joint efforts, for example, those in the Worldwide Space Station, embody the agreeable idea of room related Research and development drives.

The crossing point of neuroscience and innovation is a blossoming field of Research and development with the possibility to open the secrets of the human cerebrum and change the way we connect with machines.

Mind PC interfaces (BCIs) are gadgets that empower direct correspondence between the cerebrum and outside gadgets, opening up opportunities for neuroprosthetics, mind controlled mechanical technology, and improved mental capacities.

Progressing research in neurotechnology plans to work on the goal and productivity of BCIs, making them more open and viable for different applications. The unraveling of brain signals and the improvement of calculations for deciphering cerebrum action address key areas of examination. Research and development drives additionally investigate the moral ramifications of neurotechnology, addressing concerns connected with protection, security, and the potential for mental upgrade.

With regards to irresistible sicknesses, Research and development drives are instrumental in the improvement of immunizations, antiviral medications, and demonstrative apparatuses. The worldwide reaction to the Coronavirus pandemic embodies the direness and cooperative nature of Research and development in tending to general wellbeing emergencies. Immunization improvement, specifically, includes a mix of sub-atomic science, immunology, and clinical preliminaries to guarantee the security and viability of immunizations.

Headways in immunization advancements, like mRNA antibodies, address forward leaps with suggestions past the ongoing pandemic. Research and development drives in irresistible sickness research additionally investigate methodologies for early recognition, reconnaissance, and regulation of flare-ups. The coordination of information investigation and man-made brainpower upgrades the ability to display and anticipate the spread of irresistible sicknesses, illuminating general wellbeing mediations and techniques.

In the domain of schooling, Research and development drives center around figuring out compelling educating and learning systems, utilizing innovation for instructive upgrades, and tending to difficulties in access and value. Instructive exploration traverses different disciplines, incorporating mental science, teaching method, educational plan advancement, and the coordination of instructive innovations.

Computerized learning stages, online assets, and versatile learning advancements are regions where Research and development adds to the development of instruction. The advancement of computer generated reality (VR) and expanded reality (AR) applications in schooling opens up vivid and intuitive growth opportunities. Research on customized learning calculations expects to fit instructive substance to individual understudy needs, cultivating a more versatile and successful learning climate.

The coordination of Research and development in sociologies and humanities adds to figuring out human way of behaving, cultural elements, and social peculiarities. Research drives in brain science, social science, human studies, and related fields investigate points going from emotional well-being and social imbalance to social legacy and phonetic variety.

Headways in sociologies and humanities Research and development incorporate the utilization of computational techniques, information examination, and advanced humanities draws near. Message mining, feeling investigation, and organization examination procedures give bits of knowledge into examples of human way of behaving and collaboration. Research and development in these disciplines contributes not exclusively to scholastic information yet in addition illuminates public strategy, social mediations, and the advancement of comprehensive and fair cultural structures.

The conversation of innovative work drives would be inadequate disregarding the basic job of interdisciplinary examination. Interdisciplinary Research and development rises above customary disciplinary limits, cultivating coordinated effort between analysts with assorted aptitude. This cooperative methodology is especially significant in tending to complicated and complex difficulties that require coordinated viewpoints and arrangements.

Environmental change, for instance, requires interdisciplinary exploration that joins bits of knowledge from natural science, designing, sociologies, and strategy studies. Additionally, headways in medical care frequently result from the joint effort of scholars, PC researchers, clinicians, and architects. Interdisciplinary Research and development drives exploit the collaborations between various fields, cultivating advancement and all encompassing critical thinking.

Notwithstanding the extraordinary effect of Research and development drives across different areas, challenges continue. Subsidizing limitations, administrative obstacles, moral contemplations, and the speed of innovative change present continuous difficulties to the Research and development scene. Offsetting momentary objectives with long haul vision, guaranteeing inclusivity in research support, and tending to the expected cultural effects of arising advances are mind boggling contemplations that require smart route.

Besides, encouraging a culture of interest, innovativeness, and nonstop learning is fundamental to develop the up and coming age of scientists and pioneers. Schooling systems, mentorship programs, and cooperative stages assume critical parts in sustaining the scholarly interest and critical thinking abilities expected to drive Research and development drives forward.

All in all, the conversation of innovative work drives across different fields highlights the dynamic and interconnected nature of human information and progress.

From medical care and innovation to manageability, space investigation, and then some, Research and development drives shape the direction of our aggregate future. The cooperative endeavors of researchers, specialists, instructors, and policymakers across the globe epitomize the common obligation to propelling information, tackling complex difficulties, and making a superior, more supportable world. As we explore the outskirts of disclosure and development, the continuous exchange around Research and development fills in as a compass directing us toward a future characterized by progress, strength, and the quest for greatness.

8.3 Exploration of potential advancements and their implications for the future

The investigation of expected progressions across different spaces enlightens the interesting possibilities and groundbreaking changes that lie not too far off. From arising advances to logical leap forwards, cultural movements, and ecological developments, the direction of progress is set apart by constant investigation, revelation, and variation. This investigation divulges the potential outcomes that look for us as well as prompts thought of the significant ramifications these progressions might have for what's to come.

In the domain of man-made consciousness (artificial intelligence), the excursion of investigation leads us into a scene where AI calculations, brain organizations, and mental processing are reshaping the manner in which we communicate with innovation. The possible headways in artificial intelligence hold commitments and difficulties that wave across enterprises, social orders, and our regular routines. Improved AI capacities, powered by huge datasets and high level calculations, are ready to hoist man-made intelligence applications higher than ever.

One region where the ramifications of simulated intelligence headways are especially significant is medical care. Prescient investigation, empowered by simulated intelligence calculations, can alter clinical diagnostics, offering the capacity to conjecture potential wellbeing gambles and intercede proactively. AI models prepared on assorted patient information can give customized treatment proposals, working on the adequacy of medical services mediations. Be that as it may, the coordination of artificial intelligence in medical care likewise raises moral worries connected with protection, information security, and the potential predispositions implanted in calculations.

In the space of independent frameworks, the investigation of progressions in mechanical technology and machine independence carries us to the cusp of another period. Independent vehicles, drones, and mechanical frameworks outfitted with complex simulated intelligence are nearly changing transportation, operations, and different ventures. The possibility of completely independent vehicles exploring our streets guarantees expanded wellbeing, proficiency, and decreased gridlock. However, similarly as with any change in perspective, questions in regards to administrative structures, moral contemplations, and cultural ramifications go with these progressions.

The combination of man-made intelligence with computer generated reality (VR) and expanded reality (AR) innovations impels us into vivid and intuitive advanced domains. The expected progressions in VR and AR hold groundbreaking ramifications for fields as different as schooling, medical services, and amusement. Virtual and increased encounters offer new roads for learning, clinical preparation, and amusement, obscuring the lines between the physical and advanced universes. Notwithstanding, contemplations connected with the moral utilization of vivid advancements, their effect on psychological wellness, and the potential for making virtual partitions need cautious investigation.

Headways in environmentally friendly power advances structure a basic outskirts in the investigation of reasonable answers for address environmental change. The expected progressions in sunlight based, wind, and other environmentally friendly power sources hold the commitment of changing towards a cleaner and more maintainable energy scene. Leap forwards in energy capacity advances, like high level batteries, are key empowering agents for the boundless reception of environmentally friendly power. The ramifications of these headways stretch out past natural contemplations, affecting international elements, financial designs, and worldwide energy security.

The investigation of likely progressions in quantum registering takes us to the edge of a processing upset. Quantum PCs, utilizing the standards of quantum mechanics, can possibly tackle complex issues dramatically quicker than traditional PCs. The ramifications of quantum figuring length different fields, from cryptography and materials science to advancement issues that are as of now recalcitrant for traditional PCs. In any case, the acknowledgment of down to earth quantum PCs likewise presents specialized difficulties, including keeping up with quantum cognizance and alleviating blunders.

In the area of materials science, the investigation of potential progressions acquaints us with novel materials with phenomenal properties and applications. The advancement of metamaterials, designed to display properties not tracked down in that frame of mind, up opportunities for controlling electromagnetic waves, sound, and intensity. Uses of these progressions range from making imperceptibility shrouds to planning super proficient sensors and specialized gadgets. The ramifications for ventures like media communications, medical services, and protection are broad, proclaiming another period of material-driven development.

Hereditary designing and headways in biotechnology bring us into the unpredictable domain of controlling the structure blocks of life. The likely headways in quality altering advancements, for example, CRISPR-Cas9, offer remarkable accuracy in changing hereditary material. The ramifications of these progressions reach out from restoring hereditary sicknesses to improving harvests for improved yields and strength. In any case, the moral contemplations encompassing quality altering, including inquiries regarding unseen side-effects and the potential for fashioner children, require cautious investigation and cultural discourse.

The investigation of progressions in space investigation and stargazing moves us past our earthbound limits. Continuous missions to investigate Mars, the Moon, and far off heavenly bodies mark mankind's journey to unwind the secrets of the universe. Expected progressions in space advances, drive frameworks, and natural surroundings plans hold the commitment of extending human presence past Earth. The ramifications reach out to logical disclosure, asset usage, and the quest for a multi-planetary presence. Be that as it may, the moral contemplations of room investigation, including issues of planetary insurance and the capable utilization of extraterrestrial assets, require worldwide cooperation and moral structures.

In the domain of training, the investigation of potential headways drives us to the developing scene of advanced learning, versatile advances, and creative teaching methods. The coordination of man-made intelligence, virtual homerooms, and customized learning stages holds the possibility to upset instruction, making it more available, adaptable, and custom-made to individual necessities. The ramifications for worldwide school systems incorporate tending to disparities, cultivating deep rooted learning, and planning understudies for the difficulties of the 21st 100 years. Nonetheless, challenges connected with advanced isolates, information security, and the job of teachers in this changed scene need cautious thought.

Progressions in neurotechnology open up a boondocks in understanding and enlarging the capacities of the human cerebrum. Mind PC interfaces (BCIs) and brain prosthetics are potential progressions that hold groundbreaking ramifications for people with neurological issues, as well concerning improving mental capacities in everybody. The moral contemplations encompassing the utilization of BCIs, including inquiries of security, independence, and the potential for mental improvements, require moral systems and cultural conversations.

The investigation of likely progressions in irresistible illness research brings us into the domain of worldwide wellbeing readiness. Propels in immunization advancements, antiviral medications, and analytic apparatuses are basic for answering arising irresistible illnesses. The ramifications of these headways reach out to pandemic readiness, worldwide wellbeing value, and the cooperative endeavors expected to address worldwide wellbeing dangers. Moral contemplations, including immunization dispersion, evenhanded admittance to medicines, and the job of global collaboration in general wellbeing, structure fundamental pieces of these conversations.

Cultural movements and social changes comprise a fundamental part of the investigation of possible progressions. The coming of the Fourth Modern Upheaval, portrayed by the reconciliation of computerized advances, man-made intelligence, and the Web of Things (IoT), is reshaping economies, businesses, and the idea of work. The ramifications for work, ability advancement, and the social texture of networks are significant. Investigating likely progressions in this setting requires an all encompassing comprehension of the cultural effects, moral contemplations, and strategy systems expected to explore this extraordinary period.

The investigation of potential headways requires an intelligent assessment of the moral aspects that go with progress. Moral contemplations length assorted fields, including man-made intelligence morals, bioethics, natural morals, and the dependable utilization of arising advancements. As we explore the ramifications of expected headways, inquiries of value, protection, security, and the cultural effects of mechanical advancements need cautious and progressing investigation.

All in all, the investigation of expected headways across different spaces welcomes us to imagine a future formed by development, revelation, and flexibility. From the wildernesses of man-made intelligence and quantum figuring to the complexities of hereditary designing and the secrets of the universe, the direction of progress is set apart by ceaseless investigation and the quest for information. Exploring the ramifications of these progressions requires an aggregate and interdisciplinary exertion, including researchers, policymakers, ethicists, and society at large. As we stand at the junction of probability, the investigation of potential headways fills in as a compass, directing us towards a future that offsets mechanical advancement with moral contemplations, cultural prosperity, and the practical stewardship of our planet.

Chapter 9

Conclusion

The multifaceted embroidery of water the executives has been woven across hundreds of years, interfacing civic establishments and scenes in a cooperative dance of food. As we explore the flows of this investigation, a significant acknowledgment arises — the wonders of stream, power, and lift water system frameworks play played urgent parts in forming the shapes of human development and changing parched scenes into flourishing desert gardens. The union of old insight and current designing has birthed a tradition of pressure driven resourcefulness that rises above time and bears demonstration of the constant human quest for tackling the nurturing power of water.

Stream water system, the respected forebear of horticultural food, remains as a demonstration of the natural comprehension of water's penchant to reinvigorate the earth. The hereditary developments, from the banks of the Nile to the prolific fields of the Indus, perceived the back and forth movement of occasional waters as a musicality to be orchestrated with horticultural cycles. Tackling the regular floods, these early architects etched mind boggling organizations of waterways and barriers, diverting the restoring waters to extinguish the thirst of their harvests. In the loftiness of these old frameworks, we glimpse the beginnings of a water driven shrewdness that established the groundwork for the rural success of domains.

The authority of gravity in molding water's course turned into a fine art in itself. The shrewd plan of terraced fields, stepwells, and water systems is a demonstration of humankind's capacity to shape the scene into a material of overflow. In the consecrated domains of old Rome, the water systems stand as design wonders, huge conduits conveying the backbone of far off springs to the core of the realm. The development of the Archimedean screw, an old helical siphon, further embodies the marriage of mechanical virtuoso and water driven standards. These immortal designs reverberation the ethos of maintainability, reverberating through the hallways of history to murmur illustrations of offset and concurrence with nature.

The climb of innovation has raised the size of water system as well as enriched mankind with the influence to resist gravity itself. Lift water system, a cutting edge wonder brought into the world of designing ability, has permitted water to challenge its normal flow and climb to inundate raised territories. The unending walk of progress has seen the rise of giant siphon stations, lifting water to phenomenal levels and delivering once infertile scenes verdant. This ensemble of siphons, lines, and valves organizes an artful dance of water driven accuracy, reviving districts that were once considered unfriendly.

As we explore this broad scene of water driven wonders, it becomes clear that the development of water system frameworks reflects the direction of human resourcefulness. From the simple channels of old Mesopotamia to the transcending siphons of contemporary times, every development is a demonstration of the dauntless human soul as its continued looking for food and success. The combination of custom and innovation unfurls as a story string, winding around the past, present, and future into a consistent continuum.

The intersection of stream, command, and lift water system frameworks exemplifies more than simple mechanical accomplishments; it typifies a significant ethos of stewardship over nature's gifts. The conventional insight implanted in the stream water system frameworks from times gone past addresses a cozy comprehension of the land's heartbeat, an affirmation that water isn't just an asset however a day to day existence power to be respected. The ascendant scenes etched by gravity-driven frameworks take the stand concerning the amicable coordination of human desire with the forms of the earth. In the advanced crescendo of lift water system, we witness the sensitive harmony between mechanical ability and natural obligation, as designers explore the fragile dance among progress and manageability.

Be that as it may, this story isn't one of unrestrained achievement. It is likewise a wake up call of the unseen side-effects that can go with the control of nature's mind boggling balance. The unrestrained power of human mediation has, now and again, prompted natural disharmony, as unrestrained double-dealing of water assets has left scars on the scene. The once-great streams diminished to streams and the exhaustion of springs act as strong updates that the power of water should be tempered by an intense attention to the sensitive biological systems it supports.

In the fantastic venue of water driven wonders, value arises as a focal hero. The stream, authority, and lift water system frameworks have not generally disseminated their advantages equitably. The old trench frameworks, while feeding the fields of realms, frequently left the minimized helpless before whimsical floods and dry spells. The authority of water, whether through gravity or siphoning stations, has over and over again streamed towards the advantaged, leaving bone-dry regions grieving in disregard. The pressure driven story should be reexamined to address the basic of impartial access, guaranteeing that the nurturing remedy of water arrives at each dry corner, independent of financial limits.

In the time of environmental change, the water driven account takes on restored importance. The conventional insight encoded in stream water system, command, and gravity-driven frameworks expects a prophetic tint as humankind wrestles with the capricious dance of precipitation designs. The old comprehension of working as one with nature entices as a core value, encouraging a re-visitation of feasible practices that recognize the interconnectedness of water, land, and life. The domination of innovation, while offering uncommon conceivable outcomes, should be tempered by a guarantee to strength and versatility despite a consistently changing environment material.

The contemporary pressure driven adventure likewise unfurls against the background of a worldwide water emergency, where the interest for this limited asset exceeds its reasonable inventory. In this cauldron of shortage, the wonders of stream, authority, and lift water system frameworks become mechanical victories as well as moral objectives. The prudent administration of water assets, the saddling of water, and the reception of accuracy water system procedures arise as fundamental parts in the developing pressure driven story. The progression of development should be directed by an aggregate liability to guarantee that water, the mixture of life, is shared fairly and protected for ages yet unborn.

In the ensemble of water driven wonders, the job of administration arises as a basic guide. The progression of approaches, guidelines, and foundations shapes the forms of water the board, directing whether it overflows down the surges of manageability or streams into the void of botch. The authority of viable administration is crucial for saddle the capability of water system frameworks for everyone's best interests, guaranteeing that the advantages of mechanical advancement are not bound to the special minority. The worldwide local area remains at a junction where the harmonization of different interests, from horticulture to industry to natural preservation, becomes basic for the coordination of an economical pressure driven future.

As we close this investigation of stream, command, and lift water system wonders, the story curve follows a direction from old insight to current development, from the instinctive comprehension of nature to the saddling of mechanical ability. The pressure driven adventure is an amazing story of humankind's mission for food, thriving, and strength notwithstanding developing difficulties. A story rises above geological limits, reverberating across the scenes of civic establishments and resonating through the chronicles of time.

In the closing sections of this water powered odyssey, the source of inspiration reverberates. The domination of dependable stewardship calls, encouraging mankind to wind around a story of water the board that embraces the ethos of manageability, value, and versatility. It is a call to recalibrate the progression of progress, guaranteeing that the waters of development flood fields of overflow as well as the fields of equity and inclusivity. The pressure driven wonders, from the unassuming trenches of ancient times to the transcending siphons of the present, are not only relics of designing;

they are images of the human soul's incessant excursion to overcome the difficulties presented ordinarily.

As we stand at the conversion of past and future, the insight exemplified in the stream, domination, and lift water system frameworks welcomes us to proceed with caution on the earth, perceiving that our power is indivisible from the ascent and fall of the waters. The account of water powered wonders is very easy to read, with pages yet unwritten, sitting tight for the aggregate insight and activities of mankind to shape the parts that lie ahead. In this unfurling adventure, the narrative of water is, at last, the tale of us — a story that entices us to be the modelers of water system frameworks as well as the watchmen of the planet's most valuable asset.

9.1 Summarization of key findings and insights

In diving into the extensive domain of pressure driven wonders enveloping stream, domination, and lift water system frameworks, a union of key discoveries and bits of knowledge arises. This exhaustive investigation uncovers a story that rises above ages, winding around together the strings of old insight and present day designing. As we distil the pith of this excursion, a few urgent subjects and acknowledge come to the very front.

At its center, the adventure of pressure driven wonders is a demonstration of humankind's personal connection with water — a relationship that reaches out past simple food to the actual underpinnings of civic establishments. The stream water system frameworks of antiquated societies, from the mind boggling channels of Mesopotamia to the dazzling terraced fields of the Inca Domain, epitomize the harmonious dance among water and horticulture. These early social orders perceived the beat of occasional waters as a day to day existence force, and their imaginative water system frameworks turned into the soul of success, supporting thriving populaces and laying the foundation for social and monetary prospering.

The command of gravity in molding water's course arises as a repetitive theme since the beginning of time. From the magnificent reservoir conduits of Rome to the shrewdly planned stepwells of India, the dominance of gravity-driven frameworks remains as a demonstration of human inventiveness. These designs not just tackled the power of gravity to channel water across huge distances yet in addition etched scenes in an amicable mix of feel and usefulness. The old fashioned Archimedean screw, an early mechanical wonder, further features the cunning marriage of development and pressure driven standards, making ready for the command of water higher than ever.

The advanced time observers the zenith of water driven development with the appearance of lift water system frameworks. Here, innovation becomes the dominant focal point, empowering water to oppose gravity and arrive at heights recently considered blocked off. Monster siphon stations, outfitted with complex hardware, embody the victory of designing over geological imperatives. The command of water, once bound to the laws of nature, is currently arranged by the complex dance of siphons, lines, and valves, introducing a time where bone-dry scenes can be changed into verdant fields.

However, in the midst of the festival of mechanical power, a nuanced comprehension of potentially negative results and biological equilibrium arises. The unrestrained control of conduits, especially directly following quick industrialization and urbanization, has now and again brought about ecological debasement. Streams diminished to simple streams and springs drained past renewal act as wake up calls, helping us that the power to remember water should be tempered by a significant regard for the fragile biological systems it maintains.

Value arises as a focal subject in the pressure driven story, with verifiable water system frameworks frequently privileging the strong and leaving underestimated networks helpless before whimsical water designs. The cutting edge time requires a recalibration of this unevenness, encouraging the improvement of strategies and practices that guarantee evenhanded admittance to water assets. The power of water, whether through gravity or progressed siphoning systems, should turn into a power for civil rights, arriving at each side of the globe, paying little mind to financial standing.

The water driven adventure unfurls against the background of an evolving environment, where the consistency of precipitation designs is not generally guaranteed. The antiquated insight implanted in stream water system frameworks, with their dependence on occasional rhythms, gains recharged significance as social orders wrestle with the difficulties presented by environmental change. The command of innovation, when tackled prudently, offers versatile arrangements, from water gathering to accuracy water system, guaranteeing flexibility despite an erratic environment material.

The worldwide water emergency arises as a cauldron in the contemporary pressure driven story. As interest for water surpasses reasonable stockpile, a clarion call reverberates for dependable water the executives rehearses. The wonders of stream, power, and lift water system frameworks become mechanical accomplishments as well as moral objectives. The progression of development should be coordinated towards practical arrangements, recognizing the interconnectedness of water, land, and life. Accuracy water system, water reusing, and the tackling of elective water sources become fundamental apparatuses in exploring the maze of shortage.

Administration arises as a basic figure molding the predetermination of water assets. The progression of approaches, guidelines, and establishments turns into the imperceptible hand directing the direction of water the board. Successful administration is key for guaranteeing that the domination of water helps the aggregate great, forestalling the grouping of assets in the possession of a couple. The worldwide local area winds up at a significant point where cooperative endeavors are basic, blending different interests for the supportable coordination of water driven frameworks.

In the last sections of this water powered odyssey, the source of inspiration resonates. The domination of dependable stewardship coaxes humankind to create a story of water the executives that rises above individual and public interests. It is a solicitation to recalibrate the progression of progress, guaranteeing that the waters of advancement inundate fields of overflow as well as fields of equity, inclusivity, and natural protection. The pressure driven wonders, from old trenches to contemporary siphoning stations, fail to be simple curios of designing; they become images of the human soul's unending excursion to overcome the difficulties presented essentially.

As we stand at the conjunction of past and future, the insight typified in stream, authority, and lift water system frameworks welcomes us to proceed with caution on the earth. The account of pressure driven wonders is very easy to read, with pages yet unwritten, hanging tight for the aggregate insight and activities of humankind to shape the parts that lie ahead. In this unfurling adventure, the narrative of water is, at last, the tale of us — a story that entices us to be the engineers of water system frameworks as well as the watchmen of the planet's most valuable asset. The end sections of this pressure driven story stretch out a solicitation to every person, local area, and country to become stewards of water, guaranteeing that its command is a reference point of food, versatility, and shared thriving for a long time into the future.

9.2 Reflection on the role of lift irrigation in sustainable agriculture

Lift water system, remaining at the very front of pressure driven development, coaxes an intelligent investigation into its job in cultivating economical farming. As we dive into the subtleties of this innovative wonder, a complex embroidery unfurls, winding around together components of natural stewardship, financial practicality, and social value. The intelligent excursion into the domain of lift water system welcomes us to examine not simply the mechanical complexities of siphoning water to raised territories yet additionally the more extensive ramifications and obligations innate in its sending.

At the core of lift water system lies the daring aspiration to resist gravity and raise water to scenes recently considered unfriendly for development. This mechanical accomplishment depends on the collaboration of siphons, lines, and power sources, changing desolate landscapes into flourishing agrarian centers.

The rise of water acquaints another aspect with farming scenes, empowering the development of harvests in regions where the normal progression of water misses the mark. While the commitment of expanded farming efficiency is obviously charming, the maintainability of such intercessions requires a sensitive harmony between human necessities and natural repercussions.

One of the urgent reflections relates to the natural effect of lift water system. The power of water to higher rises, worked with by energy-escalated siphoning frameworks, brings up issues about the carbon impression related with such undertakings. The reliance on power, frequently got from customary energy sources, presents a component of natural expense. As we imagine lift water system as a foundation of maintainable farming, it becomes basic to investigate elective energy sources, for example, sun powered or wind, to control these frameworks. The progress towards sustainable power mitigates natural worries as well as adjusts lift water system to the more extensive worldwide basic of changing towards greener practices.

The natural focal point additionally reaches out to the possible disturbance of neighborhood biological systems brought about by lift water system. Adjusting the regular direction of water and acquainting it with raised landscapes might prompt potentially negative side-effects like soil disintegration, environment disturbance, and changes in neighborhood hydrology. Cautious natural effect appraisals and relief techniques should be essential to the preparation and execution of lift water system projects. Chasing manageable farming, it is basic to recognize the interconnectedness of biological systems and take a stab at mediations that improve, as opposed to corrupt, the natural texture.

One more aspect of reflection includes the monetary practicality of lift water system frameworks. While these frameworks can possibly change dry scenes into prolific fields, the financial maintainability of such tasks requires an all encompassing assessment. The underlying capital interest in setting up lift water system framework can be significant, enveloping the expense of siphons, pipelines, and related foundation. The monetary attainability is dependent upon variables like harvest determination, market elements, and the capacity of nearby networks to use the recently discovered farming potential.

The financial contemplations likewise stretch out to the functional and upkeep expenses of lift water system frameworks. The complicated apparatus engaged with siphoning water requests standard upkeep and gifted specialized help. Smallholder ranchers and neighborhood networks, frequently the essential recipients of lift water system, may confront difficulties in bearing these upkeep costs. Feasible rural practices require the underlying venture as well as a drawn out obligation to guarantee the proceeded with usefulness and proficiency of lift water system frameworks.

Social value arises as a pivotal aspect in the reflection on lift water system's part in maintainable horticulture. The advantages of upgraded water access and rural efficiency should be conveyed evenhandedly across networks. By and large, water system projects have, on occasion, exacerbated social differences, inclining toward the rich and strong while leaving underestimated populaces in a tough spot. With regards to lift water system, it becomes basic to plan approaches and execution techniques that focus on the requirements of smallholder ranchers and weak networks.

Also, the strengthening of neighborhood networks in the dynamic cycles connected with lift water system is vital. Inclusivity in arranging, execution, and the executives guarantees that the advantages of lift water system are shared justly, encouraging a feeling of pride and manageability. Local area based water system the board models, where neighborhood partners effectively partake in the administration of water assets, can act as a reference point for impartial and manageable practices.

A basic reflection likewise digs into the versatility of lift water system frameworks even with environmental change. The eccentric changes in precipitation designs, outrageous climate occasions, and changing hydrological elements present difficulties to the traditional ideal models of water the executives. Practical farming, supported by lift water system, requires versatile techniques that record for the vulnerabilities created by an evolving environment.

In this specific circumstance, the coordination of environment savvy rehearses becomes basic. The arrangement of lift water system frameworks with environment strong yields, water protection measures, and continuous weather conditions checking can improve the versatile limit of rural scenes. Besides, the joining of environment risk evaluations in the preparation and plan of lift water system projects guarantees that these frameworks are receptive to ebb and flow conditions as well as expect and adjust to future climatic vulnerabilities.

As we think about the job of lift water system in economical agribusiness, the strengthening of nearby networks and the advancement of social inclusivity arise as vital determinants. The domination of water to raised landscapes ought not be an honor restrictive to a chosen handful yet an impetus for more extensive social and monetary turn of events. Interests in limit building, schooling, and expertise improvement inside neighborhood networks add to a maintainable biological system where the advantages of lift water system are shared fairly.

The reflection on reasonable horticulture through lift water system additionally prompts an investigation of water use proficiency. Proficient water the board is principal to the manageable use of this valuable asset. Lift water system frameworks should be planned and worked with a sharp eye on improving water use, limiting wastage, and advancing practices like trickle water system and accuracy horticulture. Water shortage, exacerbated by environmental change and populace development, orders a change in perspective towards water-productive farming strategies to guarantee the life span of lift water system mediations.

The nexus between lift water system and food security surfaces as a focal thought in the intelligent talk. Feasible agribusiness, worked with by lift water system, can possibly support food creation and upgrade food security, particularly in districts inclined to water shortage. Nonetheless, this nexus requests a nuanced comprehension of food frameworks, enveloping creation as well as dissemination, access, and nourishment. The intelligent focal point extends to include the whole agri-food esteem chain, encouraging an all encompassing methodology that rises above the limits of individual rural practices.

A comprehensive reflection likewise incorporates the social and social elements of horticulture worked with by lift water system. The change of scenes through water domination isn't simply a mechanical cycle however a significant shift that resounds with social personalities and conventional information. The protection of nearby rural practices, biodiversity, and native information frameworks ought to be indispensable to the manageability story. In doing as such, lift water system turns into a scaffold between the cutting edge and the customary, encouraging a powerful balance that praises both legacy and progress.

In the more extensive setting of reasonable turn of events, lift water system coaxes us to scrutinize the polarity among rustic and metropolitan spaces. The command of water to change dry scenes challenges the regular story that places urbanization as a harbinger of progress. Manageable farming through lift water system underlines the potential for flourishing country economies, decreasing the metropolitan rustic gap, and encouraging adjusted territorial turn of events. This reflection prompts us to imagine a future where lift water system turns into an impetus for strong, self-supporting country networks.

The reflection on lift water system's job in economical horticulture finishes in a call for coordinated and versatile water the board structures. Maintainable horticulture is definitely not a disengaged attempt however an interconnected snare of environmental, social, and financial contemplations. The domination of water through lift water system frameworks should be fit with the more extensive objectives of natural preservation, civil rights, and monetary feasibility. Coordinated water assets the executives, which considers the exchange of different factors, for example, land use, environment, and financial elements, becomes basic in exploring the intricacies of practical farming.

As we finish up this intelligent excursion, the combination of experiences highlights the requirement for a change in outlook in our way to deal with water the board and rural practices. Lift water system, as an innovative wonder, offers a groundbreaking device for feasible horticulture, yet its viability relies on the prudent combination of natural, monetary, and social contemplations. The domination of water turns into a representation for the command of a comprehensive and comprehensive vision of horticulture — one that embraces flexibility, value, and versatility notwithstanding developing difficulties. In this vision, lift water system quits being just a mechanical cycle; it turns into a conductor for supporting scenes, networks, and the fragile harmony between human undertakings and the normal world.

9.3 Call to action for continued research, development, and implementation of lift irrigation systems globally.

The investigation of lift water system frameworks and their multi-layered job in reasonable farming finishes in a resonating source of inspiration. This call exudes from the acknowledgment that lift water system, with its capability to change parched scenes into useful rural center points, isn't simply a mechanical development however a foundation for tending to squeezing worldwide difficulties. The basic for proceeded with exploration, improvement, and far reaching execution of lift water system frameworks around the world is highlighted by the entwined strings of natural stewardship, monetary strengthening, social value, and environment versatility.

At the very front of this source of inspiration is the acknowledgment of the essential job lift water system frameworks play in tending to the expanding worldwide water emergency. As populaces rise, environment designs shift, and arable land reduces, the interest for economical water the executives rehearses turns out to be progressively pressing. Lift water system, by tackling the power of water to raised territories, presents a suitable answer for grow rural outskirts and upgrade water use productivity. Proceeded with investigation into the enhancement of lift water system innovations, including the improvement of energy-proficient siphons and elective power sources, is vital to opening its maximum capacity as a feasible water the board device.

The source of inspiration stretches out to the basic of incorporating lift water system frameworks into more extensive water asset the board structures. Research endeavors ought to zero in on creating models that work with the amicable concurrence of lift water system with regular biological systems, alleviating likely natural effects and advancing biodiversity. Versatile administration procedures, informed by continuous exploration, ought to direct the organization of lift water system in different natural settings, guaranteeing that these mediations upgrade, as opposed to think twice about, sensitive equilibrium of neighborhood environments.

Environment versatility remains as a foundation of the source of inspiration, underlining the requirement for continuous innovative work to sustain lift water system frameworks against the vulnerabilities fashioned by environmental change. As precipitation designs become progressively whimsical, and outrageous climate occasions present new difficulties, lift water system should advance to meet these possibilities. Exploration ought to dive into environment savvy innovations, accuracy water system rehearses, and prescient demonstrating that empower lift water system frameworks to adjust progressively to changing climatic circumstances.

The worldwide local area should embrace a cooperative soul in the turn of events and execution of lift water system frameworks. Information sharing, mechanical exchange, and limit building are vital parts of this cooperative exertion. Nations with laid out lift water system practices ought to contribute their bits of knowledge and best works on, cultivating an aggregate comprehension that rises above geological limits.

Worldwide associations, research establishments, and state run administrations ought to produce organizations to help the spread of information and aptitude, especially to areas wrestling with water shortage and food uncertainty.

Financial suitability is a key part in the source of inspiration, requiring supported innovative work endeavors to upgrade the expense viability of lift water system frameworks. The advancement of framework plan, investigation of inventive supporting models, and the mix of brilliant innovations can add to diminishing the underlying capital venture and functional expenses related with lift water system. Research tries ought to likewise investigate systems for boosting private area contribution in lift water system projects, encouraging a collaboration between monetary interests and maintainable farming practices.

Social value becomes the dominant focal point in the source of inspiration, accentuating the requirement for innovative work that guarantees the advantages of lift water system are shared fairly among assorted networks. Comprehensive administration models, participatory dynamic cycles, and local area strengthening ought to be fundamental contemplations in the plan and execution of lift water system frameworks. Exploration ought to dig into social effect appraisals, inspecting the financial elements of lift water system intercessions and distinguishing procedures to review verifiable abberations.

Instruction and mindfulness structure vital parts of the source of inspiration, perceiving that the outcome of lift water system frameworks depends on the information and commitment of neighborhood networks. Examination ought to zero in on creating instructive projects that enable ranchers with the abilities and understanding expected for compelling usage of lift water system advances. Mindfulness crusades, upheld by research discoveries, ought to disperse data about the ecological, monetary, and social advantages of lift water system, encouraging a feeling of pride and aggregate liability.

As we set out on this aggregate undertaking, the source of inspiration stresses the basic of a strategy structure that boosts and works with the broad reception of lift water system frameworks. States at the public and provincial levels ought to focus on the coordination of lift water system into their agrarian and water the executives arrangements. Examination ought to educate the advancement regarding strategy structures that line up with supportability objectives, adjusting the requirement for expanded agrarian efficiency with ecological protection and social inclusivity.

The source of inspiration additionally resounds with the basic of constant checking and assessment of lift water system projects. Examination ought to add to the advancement of powerful checking instruments that evaluate the environmental, financial, and social effects of lift water system mediations over the long run.

Versatile administration methodologies, informed by progressing research discoveries, ought to direct arrangement changes and foundation updates, guaranteeing that lift water system frameworks stay receptive to advancing circumstances.

The worldwide idea of this source of inspiration highlights the interconnectedness of water the board provokes and the common obligation to address them. Worldwide coordinated efforts, worked with by research organizations and information sharing stages, ought to encourage a cross-fertilization of thoughts and encounters. Research establishments, non-legislative associations, and legislative organizations ought to unite to make a worldwide store of best practices, contextual investigations, and mechanical developments connected with lift water system.

The source of inspiration stretches out to the instruction and preparing of another age of experts furnished with the abilities to explore the perplexing scene of lift water system and economical water the executives. Scholarly foundations ought to coordinate courses and exploration programs zeroed in on lift water system advances, ecological effect evaluations, and social elements related with water the executives. This instructive accentuation guarantees a nonstop inundation of talented people focused on propelling the outskirts of lift water system examination and execution.

Moral contemplations support the source of inspiration, underlining the obligation of specialists, policymakers, and professionals to direct their work with uprightness and a guarantee to the prosperity of both present and people in the future. Moral exploration practices ought to direct investigations into the ecological and social ramifications of lift water system, guaranteeing that the quest for information is lined up with standards of manageability, value, and equity.

All in all, the source of inspiration reverberates as an aggregate greeting to intensify endeavors, team up across boundaries, and influence the maximum capacity of lift water system frameworks chasing feasible horticulture. This is certainly not a lone undertaking however a common obligation that rises above disciplinary limits, public lines, and scholastic storehouses. It is a call to outfit the power of water as a power for good, an impetus for ecological strength, financial thriving, and social value. The continuous examination, improvement, and execution of lift water system frameworks internationally are not simply specialized tries; they are a demonstration of humankind's aggregate obligation to supporting the planet and guaranteeing a practical and prosperous future for all.

The worldwide execution of lift water system frameworks remains as a significant wilderness in tending to the complex difficulties presented by water shortage, environmental change, and the basic for maintainable horticulture. The power of water to raised landscapes, worked with by complex siphoning components, offers an extraordinary pathway to open horticultural potential in locales recently viewed as cold.

The execution of lift water system frameworks on a worldwide scale requires a nuanced approach that incorporates mechanical development, strong strategy structures, local area commitment, and a pledge to natural stewardship. As we explore this sweeping scene, it is essential to think about the different settings, difficulties, and amazing open doors that portray the worldwide undertaking to saddle the capability of lift water system.

Mechanical advancement lies at the core of the worldwide execution of lift water system frameworks. The nonstop development of siphoning advancements, materials, and energy effectiveness assumes a urgent part in improving the viability of lift water system. Innovative work endeavors should zero in on refining siphon plans, investigating elective power sources, for example, sun oriented and wind, and enhancing the general productivity of the siphoning system. Mechanical development stretches out past the actual apparatus to include brilliant water system innovations, accuracy farming, and information driven choice emotionally supportive networks. Coordination of these developments guarantees that lift water system frameworks work as water transport components as well as savvy, versatile frameworks that answer the unique requirements of horticulture and ecological circumstances.

The execution of lift water system frameworks requires a hearty strategy structure that lines up with manageability objectives, offsets financial contemplations with ecological protection, and encourages social inclusivity. States at the public and provincial levels should focus on the joining of lift water system into their water the executives and agrarian approaches. This includes the advancement of clear rules for the preparation, endorsement, and execution of lift water system projects. The approach structure ought to boost maintainable practices, energize the utilization of environmentally friendly power in lift water system activities, and advance comprehensive administration models that engage neighborhood networks.

One of the basic parts of effective execution is local area commitment. Neighborhood people group, especially those straightforwardly affected by lift water system projects, ought to be dynamic members in the dynamic cycles. Their customary information, needs, and concerns ought to be basic contemplations in project plan and arranging. Local area commitment goes past counsel; it includes engaging nearby partners to partake in the administration of water assets effectively. Local area based administration models, where neighborhood networks play a main job in the activity and upkeep of lift water system frameworks, add to a feeling of responsibility, manageability, and fair dissemination of advantages.

The worldwide execution of lift water system frameworks requires an all encompassing methodology that considers the more extensive financial elements of the districts where these frameworks are conveyed. Smallholder ranchers, frequently at the front of rural scenes, ought to be focal recipients of lift water system mediations. Exploration and execution techniques ought to zero in on understanding the necessities and difficulties looked by smallholder ranchers, fitting lift water system ventures to their particular prerequisites, and guaranteeing that the financial advantages are conveyed impartially.

Monetary components that work with admittance to credit and lessen the monetary weight on ranchers ought to be investigated to upgrade the financial practicality of lift water system for smallholder farming.

Coordinated water assets the board arises as a foundation in the fruitful execution of lift water system frameworks universally. This includes an exhaustive comprehension of the interchange between surface water and groundwater, natural contemplations, and the more extensive effects on watersheds. Watershed-level preparation and the executives techniques guarantee that the authority of water through lift water system doesn't unintentionally hurt downstream environments or compromise the accessibility of water for different clients. Research in this area ought to zero in on creating displaying apparatuses, hydrological evaluations, and best administration rehearses that work with the mix of lift water system into all encompassing water asset the executives structures.

Natural supportability is a non-debatable part of the worldwide execution of lift water system frameworks. Exploration and execution methodologies ought to focus on the preservation of normal biological systems, avoidance of soil disintegration, and the minimization of natural effects. Ecological effect evaluations ought to be vital parts of the preparation and endorsement processes for lift water system projects. Versatile administration rehearses that empower constant checking and change of lift water system tasks in light of biological changes add to an economical concurrence with regular frameworks.

In districts where water shortage is an overarching challenge, the worldwide execution of lift water system frameworks turns into a fundamental device for opening horticultural potential. Nonetheless, this execution should be joined by mindful water the board rehearses that focus on water use effectiveness. Accuracy water system procedures, like dribble and sprinkler frameworks, ought to be coordinated into lift water system ventures to limit water wastage and improve the general productivity of water conveyance. Research endeavors ought to zero in on creating advancements and practices that improve water use in lift water system, perceiving the limited idea of water assets and the basic to sensibly utilize them.

Worldwide coordinated effort and information sharing are basic in the execution of lift water system frameworks. Fruitful contextual analyses, best practices, and illustrations gained from one locale can illuminate and help others confronting comparative difficulties. Worldwide associations, research foundations, and states ought to team up to make a worldwide vault of information on lift water system. Stages for information trade, gatherings, and cooperative examination drives add to an aggregate comprehension of the complexities engaged with lift water system execution and work with a cross-fertilization of thoughts.

Limit building arises as a basic part in the worldwide execution of lift water system frameworks. This includes not just giving specialized preparing to the activity and upkeep of lift water system foundation yet in addition building institutional limits at the nearby and public levels.

Preparing projects ought to stretch out to policymakers, water chiefs, and local area pioneers, guaranteeing that there is a mutual perspective of the intricacies and valuable open doors related with lift water system. Limit building drives ought to be custom fitted to the particular necessities of every district, perceiving the assorted settings wherein lift water system is carried out around the world.

Interest in innovative work is principal for the supported outcome of lift water system frameworks universally. The powerful idea of natural circumstances, mechanical headways, and financial elements requires nonstop advancement and transformation. States, worldwide associations, and confidential area elements ought to allot assets for research drives that investigate new boondocks in lift water system innovation, ecological administration, and financial elements. These ventures not just add to the refinement of existing practices yet additionally make ready for the improvement of state of the art arrangements that address arising difficulties.

Observing and assessment structures are crucial in the worldwide execution of lift water system frameworks. Thorough appraisals of the ecological, financial, and social effects of lift water system mediations add to versatile administration rehearses. Long haul checking drives ought to be laid out to follow the presentation of lift water system frameworks, assess their flexibility even with evolving conditions, and illuminate strategy changes. Research in this space ought to zero in on creating normalized pointers, appraisal strategies, and effect assessment devices that work with a complete comprehension of the results of lift water system projects.

All in all, the worldwide execution of lift water system frameworks addresses a groundbreaking pathway to address the perplexing difficulties of water shortage, environmental change, and feasible farming. This try requires a comprehensive methodology that includes mechanical development, hearty strategy structures, local area commitment, and a promise to natural stewardship. The rising of water to raised landscapes, worked with by lift water system, should be directed by the standards of value, versatility, and dependable asset the board. As countries and networks across the globe wrestle with the basic to improve rural efficiency while protecting the climate, the execution of lift water system frameworks remains as a guide of development and trust.

The worldwide execution of cutting edge water system frameworks, especially the far and wide reception of lift water system, addresses a basic boondocks in tending to the steadily heightening difficulties presented by water shortage, environmental change, and the journey for feasible horticulture. As we leave on this sweeping undertaking, it is fundamental to dig into the diverse aspects that portray the execution of such frameworks on a worldwide scale. This excursion includes exploring through the complicated snare of mechanical progressions, strategy systems, local area commitment, ecological stewardship, and the basic for versatile, versatile practices.

At the center of the worldwide execution lies the headway of water system innovations, with lift water system frameworks expecting a noticeable job. These frameworks, driven by the authority of water through unpredictable siphoning instruments, act as extraordinary devices to defeat geological limitations and develop bone-dry scenes. Mechanical development turns into a key part in this worldwide pursuit, requesting persistent innovative work endeavors to refine siphon plans, investigate elective energy sources, and upgrade by and large framework productivity. The development stretches out past simple mechanical complexities to incorporate savvy water system advances, accuracy horticulture practices, and information driven choice emotionally supportive networks. A worldwide obligation to innovative progression guarantees that lift water system frameworks develop into clever, versatile organizations equipped for answering progressively to the developing necessities of horticulture and the climate.

Simultaneous with mechanical headways, the worldwide execution requires a strong strategy system that resounds with supportability goals, offsets financial contemplations with ecological preservation, and encourages social inclusivity. States at public and provincial levels bear the obligation of incorporating lift water system into their water the executives and horticultural strategies. This includes the advancement of clear rules for project arranging, endorsement, and execution. The arrangement structure ought to boost reasonable practices, energize the utilization of environmentally friendly power in lift water system activities, and advance comprehensive administration models that engage neighborhood networks. Viable strategy combination turns into a foundation, giving the vital platform to the boundless execution of lift water system frameworks.

Local area commitment arises as a basic feature of effective execution. Neighborhood people group, especially those straightforwardly affected by lift water system projects, should rise above the job of simple partners to become dynamic members in dynamic cycles. Their customary information, needs, and concerns ought to be vital contemplations in project plan and arranging, encouraging a feeling of pride and shared liability. Past discussion, local area commitment includes enabling nearby partners to partake in the administration of water assets effectively. Local area based administration models, where neighborhood networks play a main job in the activity and upkeep of lift water system frameworks, add to manageability and the impartial conveyance of advantages.

The worldwide execution of lift water system frameworks requires an all encompassing methodology that considers the more extensive financial elements of the districts where these frameworks are sent. Smallholder ranchers, frequently the foundation of agrarian scenes, ought to be focal recipients of lift water system mediations. Examination and execution procedures ought to zero in on understanding the necessities and difficulties looked by smallholder ranchers, fitting lift water system ventures to their particular prerequisites, and guaranteeing that monetary advantages are conveyed fairly. Monetary systems working with admittance to credit and diminishing the monetary weight on ranchers ought to be investigated to upgrade the financial reasonability of lift water system for smallholder agribusiness.

Coordinated water assets the executives arises as a foundation in the effective execution of lift water system frameworks universally. This includes a thorough comprehension of the transaction between surface water and groundwater, biological contemplations, and the more extensive effects on watersheds. Watershed-level preparation and the board methodologies guarantee that the power of water through lift water system doesn't incidentally hurt downstream environments or compromise water accessibility for different clients. Research in this space ought to zero in on creating demonstrating devices, hydrological evaluations, and best administration rehearses that work with the mix of lift water system into all encompassing water asset the board structures.

Ecological supportability is a non-debatable part of the worldwide execution of lift water system frameworks. Examination and execution methodologies ought to focus on the protection of regular biological systems, counteraction of soil disintegration, and the minimization of natural effects. Ecological effect evaluations ought to be vital parts of the preparation and endorsement processes for lift water system projects. Versatile administration rehearses that empower continuous observing and change of lift water system tasks because of biological changes add to a manageable concurrence with regular frameworks.

In locales wrestling with water shortage, the worldwide execution of lift water system frameworks turns into a vital technique for opening farming potential. In any case, this execution should be joined by mindful water the executives rehearses that focus on water use effectiveness. Accuracy water system strategies, like trickle and sprinkler frameworks, ought to be incorporated into lift water system tasks to limit water wastage and upgrade the general effectiveness of water dissemination. Research endeavors ought to zero in on creating advances and practices that streamline water use in lift water system, perceiving the limited idea of water assets and the basic to prudently utilize them.

Worldwide joint effort and information sharing are basic in the execution of lift water system frameworks. Effective contextual analyses, best practices, and examples gained from one locale can illuminate and help others confronting comparable difficulties. Worldwide associations, research foundations, and states ought to team up to make a worldwide storehouse of information on lift water system. Stages for information trade, gatherings, and cooperative exploration drives add to an aggregate comprehension of the complexities engaged with lift water system execution and work with a cross-fertilization of thoughts.

Limit building arises as a basic part in the worldwide execution of lift water system frameworks. This includes not just giving specialized preparing to the activity and up-keep of lift water system foundation yet in addition building institutional limits at the nearby and public levels.

Preparing projects ought to reach out to policymakers, water directors, and local area pioneers, guaranteeing that there is a mutual perspective of the intricacies and valuable open doors related with lift water system. Limit building drives ought to be customized to the particular necessities of every locale, perceiving the different settings wherein lift water system is carried out around the world.

Interest in innovative work is principal for the supported progress of lift water system frameworks worldwide. The powerful idea of natural circumstances, mechanical headways, and financial elements requires consistent development and variation. State run administrations, worldwide associations, and confidential area substances ought to apportion assets for research drives that investigate new outskirts in lift water system innovation, natural administration, and financial elements. These ventures not just add to the refinement of existing practices yet in addition make ready for the improvement of state of the art arrangements that address arising difficulties.

Checking and assessment structures are key in the worldwide execution of lift water system frameworks. Thorough evaluations of the ecological, financial, and social effects of lift water system mediations add to versatile administration rehearses. Long haul observing drives ought to be laid out to follow the presentation of lift water system frameworks, assess their strength notwithstanding evolving conditions, and illuminate strategy changes. Research in this space ought to zero in on creating normalized pointers, evaluation strategies, and effect assessment devices that work with a complete comprehension of the results of lift water system projects.

All in all, the worldwide execution of lift water system frameworks addresses a groundbreaking pathway to address the complicated difficulties of water shortage, environmental change, and economical horticulture. This try requires a comprehensive methodology that incorporates mechanical development, powerful strategy structures, local area commitment, and a promise to natural stewardship. The domination of water to raised territories, worked with by lift water system, should be directed by the standards of value, versatility, and capable asset the board. As countries and networks across the globe wrestle with the basic to upgrade rural efficiency while defending the climate, the execution of lift water system frameworks remains as a signal of development and trust.

www.ingramcontent.com/pod-product-compliance
Lightning Source LLC
LaVergne TN
LVHW050628200726
843506LV00010B/1160